Climate Change

by Elizabeth May *and* John Kidder

Zoë Caron
Co-author of *Global Warming For Dummies*

for dummies®
A Wiley Brand

Climate Change For Dummies®

Published by: **John Wiley & Sons, Inc.,** 111 River Street, Hoboken, NJ 07030-5774, www.wiley.com

Copyright © 2022 by John Wiley & Sons, Inc., Hoboken, New Jersey

Published simultaneously in Canada

For general information on our other products and services, please contact our Customer Care Department within the U.S. at 877-762-2974, outside the U.S. at 317-572-3993, or fax 317-572-4002. For technical support, please visit https://hub.wiley.com/community/support/dummies.

Wiley publishes in a variety of print and electronic formats and by print-on-demand. Some material included with standard print versions of this book may not be included in e-books or in print-on-demand. If this book refers to media such as a CD or DVD that is not included in the version you purchased, you may download this material at http://booksupport.wiley.com. For more information about Wiley products, visit www.wiley.com.

Library of Congress Control Number: 2022933663

ISBN: 978-1-119-70310-5

ISBN 978-1-119-70317-4 (ebk); ISBN 978-1-119-70312-0 (ebk)

SKY10033635_031622

Contents at a Glance

Table of Contents

Introduction

On Monday, news reports tell you the ice caps are melting, and people everywhere are about to be swept off in a giant flood. On Tuesday, you hear a radio interview with a scientist who says global warming is galloping faster than expected. Wednesday finds you standing in the grocery line, listening to people upset to see deaths in the unprecedented heatwave. By Thursday, you just don't know whether it's time to actively dig in to be engaged in the issue.

Think of today as Friday — the day all these stray pieces come together right here in your hands, thanks to *Climate Change For Dummies*.

Climate change is here. It's no longer a future threat — catastrophic events due to global warming occur somewhere around the world daily. Wildfires, extreme droughts, heatwaves, more intense hurricanes, and deadlier tornados are causing massive economic losses and tragic loss of lives.

We don't know how bad things can get; we only know that humanity has time to avoid the worst. The more climate changes around the world, the more you have to understand what global warming is. But you know what? It's really quite exciting. Although global warming is connected to scary scenarios featuring soaring temperatures and worsening hurricanes and monsoons, it's also a link to a better future. Global warming is opening doors for the development of new types of energy, leading the shift to reliable energy sources, and creating a vision of a greener tomorrow. And the best part? You're right in the middle of it all, helping to make those changes.

About This Book

Climate Change For Dummies is your guide to climate change. We use the terms *climate change, climate emergency, climate crisis,* and *global warming* interchangeably in this book, though they're slightly different things, as we discuss in Chapter 1. This book gives you the basics so that you can understand the problem, relate it to your daily life, and be inspired to start working on solutions to this complex and important issue.

In this book, we explain the concepts behind global warming clearly and simply by using the latest, most credible science, mainly from the Sixth Assessment Report of the Intergovernmental Panel on Climate Change (IPCC). The IPCC is a team of more than 2,000 scientists who assess peer-reviewed climate change science and compile the assessments into a number of reports. These reports are mainly to inform the politicians and bureaucrats at the United Nations' decision-making table, but anyone looking for detailed scientific information on climate change can read them. The IPCC is the most credible source of climate change information in the world today. (We discuss the IPCC in greater detail in Chapter 11.)

Although this book covers what climate change is and its impact on the world, *Climate Change For Dummies* isn't just about the science. The handy guide also looks at a wide range of solutions to tackle climate change. We explore everything from the big-picture solutions that governments and businesses can implement to a slew of practical, can-do-it-today solutions for you at work, at home, and on the road.

In this book we include the following updates:

>> The increased urgency of acting to hold to 1.5 degrees C

>> The Paris Agreement

>> The significant and growing threat of ocean acidification

>> The good news of rapid acceleration of renewable energy

Foolish Assumptions

We wrote this book assuming that you know zero, nil, zilch about global warming. You don't have to look up the definitions of big, ridiculous words or drag out your high school science textbook to read this book.

We also assume, however, that you know climate change exists, that you recognize humans contribute to this problem, and that you want to understand why global warming is happening.

Icons Used in This Book

Throughout this book, you see little icons sprinkled in the left margin. These handy symbols flag content that's of particular interest.

This icon marks feel-good stories and major advances in the fight against climate change.

This icon marks a piece of information that's important to know in order to understand global warming and the issues that surround it.

Don't worry about reading paragraphs with this icon. This icon flags material that we think is interesting, but might be a little too detailed for your tastes.

Ready to make a difference? This icon points you to simple solutions that can help you reduce your greenhouse gas emissions or become a part of a bigger solution.

This icon marks paragraphs in which we talk about serious issues that humanity needs to deal with as soon as possible.

Beyond This Book

This book is full of information in plain English about climate change. If you want some additional pieces that you can refer to on a regular basis, check out the book's Cheat Sheet at www.dummies.com. Just search for "Climate Change For Dummies Cheat Sheet."

Where to Go from Here

This book like all other Dummies books is linear, meaning you can read any chapter or section that interest you. If you already know something about the subject or want to find out more about a specific topic, you can open any chapter and start reading. You also can scan the Table of Contents or the index, find a topic that piques your interest, and turn to that chapter to begin reading.

If you're entirely new to the subject of climate change, you'll likely want to read this book the old-fashioned way, starting at the beginning and working through to the end. Or, if you're interested in reading about potential solutions, head to Part 5. No matter where you start, you can find out about causes, effects, progress, and probable solutions.

1

Understanding Climate Change

Chapter **1**

Covering the Basics of Climate Change

The phrase "global warming" has been in the news since the late 1980s, but climate change, as global warming is also known, has been around much longer. In fact, it has been a constant throughout history. Earth's climate today is very different from what it was 2 million years ago, let alone 10,000 years ago. Since the beginnings of the most primitive life forms, this planet has seen many different climates, from the hot, dry Jurassic period of the dinosaurs to the bleak, frozen landscapes of the ice ages.

Today, however, the planet is experiencing something new: Its climate is experiencing rapid and dangerous changes. Scientists are certain that these changes have been caused by emissions produced by human activities. By examining previous changes in the Earth's climate, using computer models, and measuring current changes in atmospheric chemistry, they can estimate what global warming might mean for the planet, and their projections are scary.

Fortunately, Earth isn't locked into the worst-case-scenario fate yet. By banding together, people can put the brakes on global warming. In 2009, when this book was first released, we had more time to apply the brakes than now. This chapter explains the essentials of global warming and what everyone can do to achieve a greener future.

Getting a Basic Overview — Global Warming 101

When "global warming" became a household phrase, greenhouse gases *(GHGs)*, which trap heat in the Earth's atmosphere, got a bad reputation. After all, those gases are to blame for heating up the planet. But, as we discuss in Chapter 2, GHGs in reasonable quantities aren't villains, they're heroes. They capture the sun's warmth and keep it around so that life as it's known is possible on Earth. The problem starts when the atmosphere contains too great an amount of GHGs. (In Chapter 3, we look at how scientists have determined the correlation between carbon dioxide in the atmosphere and temperature.)

Other factors, which we discuss briefly in the following sections, affect the Earth's climate. Some are short-term — mostly those are seen as variations in weather, like El Niño or El Niña. The ones that matter most, though, are those that have long-term effects on climate. When the overall temperature of the Earth and the oceans rise, that's not just a change in the weather. And it's not just a normal variation that might have been observed in the past. That's a change in the Earth's climate.

Heating things up with GHGs

Human activities — primarily, the burning of fossil fuels (which we look at in the section "Tapping the Roots of Global Warming," later in this chapter) — have resulted in growing concentrations of carbon dioxide and other GHGs in the atmosphere. As we explain in Chapter 2, these increasing quantities of GHGs are retaining more and more of the sun's heat. The heat trapped by the carbon dioxide blanket is raising temperatures all over the world — hence, *global warming*.

Since the beginning of the Industrial Revolution, Earth has seen a 1.4-degree Fahrenheit (1.1 degree Celsius) increase in global average temperature because of increased GHGs in the atmosphere. Temperatures in polar regions, such as the Arctic, are experiencing temperature rises that are three times the global average.

Investigating other causes of global warming

Global warming is a very complex issue that you can't totally understand without looking at the ifs, ands, or buts. Scientists have been certain for decades that the rapid changes to climate systems are due to the buildup of GHGs. With every new scientific report, they're more certain and more concerned that changes must be made to avoid the worst-case scenarios. Other elements play a role in shaping the planet's climate, however, including the following:

>> **Cloud cover:** Clouds are connected to humidity, temperature, and rainfall. When temperatures change, so does the cloud cover — and vice versa.

>> **Long-term climate trends:** The Earth has a history of going in and out of ice ages and warm periods. Scientific records of carbon dioxide levels in the atmosphere go back 800,000 years, but people can only give educated guesses about the climate earlier than that.

>> **Solar cycles:** The sun goes through a cycle that brings it closer to or farther away from the Earth. This cycle ultimately affects the temperature of this planet and thus the climate. However, scientists have eliminated solar cycles as a factor in current warming.

We go over these other issues in greater detail in Chapter 3.

CLIMATE CHANGE — THE STORY IN A NUTSHELL

Earth has been around for about 5 billion years, starting as a ball of swirling gas and dust left over from the formation of the sun. In the first part of this very long time, the iron and silica that make up most of the planet separated — the hot heavy iron went down to the core and the lighter silicates came to the surface and cooled. Volcanoes belched material and gases up to the surface. Continents formed and move around on the surface of the planet. The Earth froze from pole to pole, heated, thawed, froze again. The mix of gases in the atmosphere changed as volcanoes and sun had their effects.

(continued)

(continued)

An overview of life on Earth

Life began and then ebbs and flows ensued:

- **3½ billion years ago:** Single-celled organisms and viruses appeared.

- **2½ billion years ago:** Photosynthesis began in bacteria; sunlight provided the energy to convert carbon to cellular growth and emit oxygen as waste.

- **900 million years ago:** The first multi-celled organisms appeared.

- **450 to 600 million years ago:** Life exploded, and plants and animals from the oceans began to colonize the land.

- **250 million years ago:** The first mass extinction happens — the survivors are the early dinosaurs and mammals.

- **200 million years ago:** Another mass extinction occurs — now the dinosaurs become dominant. At the same time, some little mammals become warm-blooded, with new abilities to live in varying climates.

- **150 to 100 million years ago:** The first birds and flowering plants appear; large dinosaurs coexist with four distinct groups of mammals.

- **66 million years ago:** An asteroid hits eastern Mexico, the cloud of dust and steam blocks the sun for years, plants die, and the dinosaurs (and all other animals weighing more than 55 pounds [25 kg]) go extinct.

- **55 million years ago:** Another mass extinction happens, this one perhaps caused by a rise in greenhouse gases, that make the atmosphere a more effective insulator and causes Earth to heat past the survival limits of many species. It's a tough place to live, Earth. Nothing is certain.

- **6 million years ago:** The first humans appear.

Human beings have been around in the same basic form for 6 million of the 5 billion years of Earth, one-eighth of one percent of all that time. During that (short) time, humans survived ice ages and developed tools, agriculture, writing, states and governments, music, and art. The human population grew constantly but slowly, held within the limits of what Earth and natural processes could provide, at about 0.04 percent per year, from 10,000 BC to 1700 AD. By 1700 about 600 million people lived on Earth, rising to about 1 billion by 1800.

But then things changed. Between 1800 and 1928, the human population doubled to 2 billion. From then on, the rate of increase rose rapidly until about 1968 — and population went up to 2.5 billion, to 5 billion by 1987, and 7.7 billion by 2019. The rate of increase peaked in 1968 and has been decreasing ever since, but still the population is expected to rise to a maximum of about 11 billion by 2100.

So humans have come to dominate Earth as no other life form ever has. And it's not just people — the animals that humans keep are now by far the largest part of the world's total animal biomass (*biomass* is the total mass of living matter in a given area).

In addition to the rising population, humans learned in the early part of the 1800s how to use the energy stored in the Earth millions of years ago. It all came from those old plants that millions of years ago used energy from the sun to grow and make their carbon tissue. When the plants died, their tissues rotted and decomposed, and over millions of years were compressed into coal and oil.

Beginning to use fossil fuels

Black or brown coal, the compressed remains of ancient plants, is a wonderful source of high-density energy. It's sometimes easy to find on the surface of the Earth, so it has been used for thousands of years for fuel (and humans had learned to make a sort of coal equivalent, charcoal, by heating wood without enough oxygen to actually burn). But most of the coal in the world is underground, not so easy to pick up and take home to the fire. People dug shallow mines to get the coal out, but often water would flood in and prevent further digging.

And then came the revolution, the start of what this book is about.

In the late 1700s and early 1800s, a Scottish engineer named James Watt made the first steam engines with high enough efficiency to be used to reliably pump water out of coal mines. So the coal provided the fuel to heat the water to make the steam to drive the engine to work the pump to get rid of the water to get at the coal, and the Industrial Revolution was launched.

With abundant coal, industrial applications became possible all over England. Engines made by Watts and others drove all sorts of processes, factories of all sorts, wool and cotton spinning, steam looms, steel mills, railroads and steamships: Coal powered England to become the first great industrial empire. The technology was then exported around the world, to Europe and the new states in America, and industrial output exploded.

(continued)

(continued)

With this new kind of industry, people began to be employed in large numbers in centralized locations. The move from the country to the cities accelerated. In the cities, coal was burned for heat, for hot water, and "coal gas" lit indoor and outdoor spaces. People worked longer hours of work and enjoyed evening entertainments in theaters and music halls.

But wait. There's still more.

The second Industrial Revolution begins

In 1859, oil was produced from a well in Pennsylvania, and the second Industrial Revolution began. Coal remained dominant for a long time, but the use of oil and its companion product methane gas (called *natural gas* to help with marketing) grew rapidly until the use of oil equaled coal by the 1950s and then displaced coal from most uses (except to make electricity and steel) by the 1970s. Oil is easier to handle than coal, produces more usable energy with less smoke and soot, and is just a better fuel source for railroads and ships, for industry, and for electrical generation. So King Coal lost its crown.

But humans were making more and more stuff and were still burning a lot of coal and now a lot of oil and gas as well. When that coal and oil was burned, humans got to use that ancient solar energy again to make things and move things and keep things warm. And all that carbon was released again, off into atmosphere as carbon dioxide and other gases called *greenhouse gases* because they act like the glass in a greenhouse to keep heat in. So, like any good greenhouse, the Earth became warmer. And it's still getting warmer today. And that's the problem to solve.

Tapping into The Roots of Global Warming

Just what are humans doing to release all those GHGs into the atmosphere? You can pin the blame on two main offenses, which we discuss in the following sections: burning fossil fuels and deforestation.

Fueling global warming

When you burn *fossil fuels*, such as coal and oil (named fossil fuels because they're composed of ancient plant and animal material), they release vast amounts of GHGs (largely, but not exclusively, carbon dioxide), which trap heat in the atmosphere. Fossil fuels are also a limited resource — meaning that humanity can't count on them over the long term because eventually they'll just run out.

The fossil fuel that produces the most GHG emissions is coal, and burning coal to produce electricity is the major source of coal-related GHGs. The second-worst offender is using gasoline and diesel for transportation, followed by burning oil to generate heat and electricity. In fact, if people could replace the coal-fired power plants around the world and switch away from the internal combustion engine, humanity would have most of the problem licked. This switch is happening now, more and more quickly, but industries that have been built on the fossil fuel bonanza, and their supportive governments and bankers, continue to delay the inevitable progress. (Check out Chapter 4 for more fossil fuel info, Chapter 13 for the scoop on energy alternatives, and Chapter 17 for an introduction to the disruption expected and feared by those industries.)

Heating up over deforestation

Forests, conserved land, and natural habitats aren't important just for the sake of saving trees and animals. Forests and all greenery are important players in keeping the climate in check. Plants take in the carbon that's in the atmosphere and give back oxygen, and older trees hold on to that carbon, storing it for the duration of their lives. By taking in carbon dioxide, they're significantly reducing the greenhouse effect. (See Chapter 2 for more about how plants help the Earth keep atmospheric carbon at a reasonable level.)

Unfortunately, much of the world's forests have been cut down to make way for farmland, highways, and cities. Deforestation is responsible for about a quarter of GHG emissions. Rainforests and mangrove forests (very productive forests that grow in wetlands) are especially good at soaking up carbon dioxide because they breathe all year round. Temperate forests, on the other hand, don't absorb much carbon dioxide over the winter, practically going into hibernation. (Chapter 5 has more about deforestation.)

Examining the Effects of Global Warming around the World

This book could easily be called *Climate Emergency For Dummies*. Although "global warming" is the common term for the climate changes that the planet's experiencing (and scientists agree that average global temperature is increasing with the buildup of GHGs), the term doesn't tell the whole story. The Earth's average surface temperature is certainly going up. But while the average keeps rising, the variations around the average are also getting larger and larger. So some areas of the planet may actually get colder or experience more extreme bouts of rain, snow, or ice build-up. Consequently, most scientists prefer the term "climate change." In the following sections, we look at how different places around the world will experience climate change.

WARNING

Much of this section is pretty depressing. But nothing is exaggerated — the information here is all based on peer-reviewed scientific reports. Just how serious could the global impact of climate change be? The first global comprehensive scientific conference, which was held in Toronto, Canada, in 1988, described the potential effects of climate change this way: "Humanity is conducting an unintended, uncontrolled, globally pervasive experiment whose ultimate consequences could be second only to a global nuclear war."

Of course, different parts of Earth have very different climates now, and climate change won't affect every part of the planet in the same way. The following sections explain in general terms how some parts of the world are being affected by climate change.

The United States and Canada

In the United States and Canada, average temperatures have been rising because of climate change. As a result, the growing season has lengthened; trees have been sucking in more carbon, and for a while, farms were more productive. The recent years have had far more severe wet years followed by extreme drought. The 2021 drought had negative and long-term impact on wheat, corn, and other crops, according to *Forbes*.

Many plants and animals are spreading farther north to adapt to climate changes, affecting the existing species in the areas to which they're moving. Increased temperatures have already been a factor in more forest fires and wildfires and damage by forest insects, such as the pine beetle epidemic in the interior of British Columbia, Canada. (See Chapter 8 for more information about how global warming will affect animals and forests.)

Scientists project that the United States and Canada will feel the effects of climate change more adversely in the coming years. Here are some of the problems, anticipated to only get worse if civilization doesn't dramatically reduce GHG emissions:

>> **Droughts and heat domes:** Rising temperatures are increasing droughts in areas that are already arid, putting even larger pressure on scarce water sources in areas such as the U.S. Southwest. In Canada, 600 people died from extreme heat in 2021 as a *heat dome* (happens when the atmosphere traps hot ocean air) formed over the west and drove temperatures to record levels.

Some areas of British Columbia experienced temperatures of 122 degrees Fahrenheit (50 degrees Celsius).

>> **Evaporating lakes:** The cities in the great heartland of the Great Lakes Basin will face retreating shorelines when the water levels of the Great Lakes drop because of increased evaporation. Lower water levels will also affect ship and barge traffic along the Mississippi, St. Lawrence, and other major rivers.

>> **Floods:** Warmer air contains more moisture, and North Americans are already experiencing more sudden deluge events, causing washed out roads and bridges, and flooded basements and even Manhattan's subways. In British Columbia in 2021, a form of rainstorm so extreme it's called an *atmospheric river* caused massive flooding. The estimated damage to farms and transportation infrastructure was about $5.9 billion US, $7.5 billion Canadian. Bridges and other sections of roads and highways were washed away, isolating coastal areas from the rest of Canada for weeks.

>> **Major storms:** Warming oceans increase the risk of extreme weather that will plague coastal cities. Think of Hurricane Katrina, arguably the most devastating weather event ever to hit a North American city, as a precursor of storms like Superstorm Sandy. Katrina was whipped into a hurricane with a massive punch from the super-heated waters of the Gulf of Mexico in 2005. In 2021 a devastating series of tornadoes, way outside the "normal" tornado season, clobbered the southern and central United States, killing almost 100 people and causing millions in damage.

REMEMBER

Not all extreme weather events are hurricanes. Global warming is expected to increase ice storms in some areas and thunderstorms in others.

>> **Melting glaciers:** Glaciers from the Rockies to Greenland, are in rapid retreat, according to the National Snow and Ice Data Center. Glacier National Park could someday be a park where the only glacier is in the name. When glaciers go, so does the spring recharge that flows down into the valleys, increasing the pressure on the remaining water supplies. People who depend on drinking water from rivers or lakes that are fed by mountain glaciers will also be vulnerable.

>> **Rising sea levels:** Water expands when it gets warmer, so as global average temperatures rise, warmer air warms the ocean. Oceans are expanding, and sea levels are rising around the world, threatening coastal cities — many of which are in the United States and Canada. This sea level rise will be far more devastating if ice sheets in Greenland and Antarctica collapse.

Changes across northern Canada and Alaska are more profound than in the south. We discuss these impacts in the section "Polar regions," later in this chapter.

On average, North Americans have many resources, in comparison to developing regions of the world, to help them adapt to climate change. The Intergovernmental Panel on Climate Change (IPCC) says Canada and the United States can take steps to avoid many of the costs of climate change, to better absorb the effects, and to avoid the loss of human lives. For example, North America could establish better storm warning systems and community support to make sure that poor people in inner cities have some hope of relief during more frequent killer heat waves. (See Chapter 10 for more information about what governments can do to help their countries adapt to the effects of climate change.)

Latin America

South America has seen some strange weather in the past few years. Drought hit the Amazon in 2005, Bolivia had hail storms in 2002, and the torrential rainfalls lashed Venezuela in 1999 and 2005. In 2003, for the first time ever, a hurricane hit Brazil. More recently, the World Meteorological Association says:

> "Latin America and the Caribbean (LAC) is among the regions most challenged by extreme hydro-meteorological events. This was highlighted in 2020 by the death and devastation from Hurricane Eta and Iota in Guatemala, Honduras, Nicaragua, and Costa Rica, and the intense drought and unusual fire season in the Pantanal region of Brazil, Bolivia, Paraguay, and Argentina. Notable impacts included water and energy-related shortages, agricultural losses, displacement and compromised health and safety, all compounding challenges from the COVID-19 pandemic."

Other changes in Latin America may be attributable to global warming. Rain patterns have been changing significantly. More rain is falling in some places, such as Brazil, and less in others, such as southern Peru. Glaciers in the Andes Mountains and across the continent are melting. This glacier loss is a particular problem in Bolivia, Ecuador, and Peru, where many people depend on glacier-fed streams and rivers for drinking water and electricity from small-scale hydroelectric plants. (See Chapter 9 for more about how global warming will affect humans.)

Scientists project that the worst is yet to come. The IPCC models anticipate that about half of the farmland in South America could become more desert-like or suffer saltwater intrusions. If sea levels continue to rise at a rate of 0.08 to 0.12 inches (2 to 3 millimeters) per year, it could affect drinking water on the west coast of Costa Rica, shoreline tourism in Mexico, and mangroves in Brazil.

The threat to the Amazonian rainforest from logging and burning has attracted the concern of celebrities such as Sting and Leonardo DiCaprio. But human-caused global warming could potentially do more damage than loggers. By mid-century, the IPCC predicts that parts of the Amazon could change from wet forest to dry grassland, and that reduction in rainfall during dry months will reduce agricultural yields. Recent scientific reports confirm even a 2 degree C temperature increase could wipe out the Amazon. (We cover how ecosystems will be affected by climate change in Chapter 8.)

Europe

Recent findings have shown that climate change is already well under way in Europe. Years ago, the IPCC projected the changes that the continent is experiencing today: rising temperatures, devastating floods, increased intensity and frequency of heat waves, and increased glacier melt.

As for what's in store for Europe, the IPCC reports a 99-percent chance that Europe will experience other unfavorable climate changes. Changes experienced so far include the following:

» **More flash floods and loss of life in inland areas:** In 2021, floods in Germany and Belgium killed more than 200 people and caused billions in damage — experts agree that such previously called "once in 400-year" floods are much more likely because of climate change.

» **More heat waves, forest fires and droughts in central, eastern, and southern Europe:** These events significantly impacted health and tourism in southern Europe in particular. The worst year on record for forest fires was 2019, until 2021 burned 1.2 million acres (half a million hectares). Much of the forest burned was in southern Europe, but fires are having increasing effects in the north as well.

» **Rising sea levels, which will increase erosion:** These rising sea levels, coupled with storm surges, will also cause coastal flooding. The Netherlands and Venice are experiencing greater impacts than other areas in Europe dealing with the rising sea level. Venice, a 1,600-year-old Italian city that is one of the world's greatest heritage sites, is built on log piles (which are gradually sinking) among canals, and so is particularly vulnerable to climate change. Rising sea levels are increasing the frequency of high tides that inundate the city.

A report published by the U.S. National Academy of Science says that loss of up to 50 percent of Europe's native species of plants and animals may be likely if climate change isn't arrested. Fisheries will also be stressed.

These impacts are all serious, but none of them represents the worst-case scenario — the Gulf Stream stalling. The results of this (stopping of a major ocean current) would be disastrous for Europe. (We look at the Gulf Stream Ocean current issue in Chapter 7.)

Africa

On a per-person basis, Africans have contributed the very least to global warming because of overall low levels of industrial development. Just look at a composite photo of the planet at night: The United States, southern Canada, and Europe are lit up like Christmas trees, burning energy that results in GHG emissions. Africa, on the other hand, shows very few lights: some offshore oil rigs twinkle, and a few cities shine, but the continent is mostly dark.

Despite contributing very little to the source of the problem, many countries in Africa are already experiencing effects of global warming. East Africa Hazards Watch says

> "Major cities in East Africa have witnessed an increase in temperatures that almost doubles the 1.1 degrees C warming that the globe has experienced since pre-industrial times. Since 1860 Addis Ababa (Ethiopia) has warmed by 2.2 degrees C, Khartoum (Sudan) by 2.09 degrees C, Dar es Salaam (Tanzania) by 1.9 degrees C, Mogadishu (Somalia) by 1.9 degrees C, and Nairobi (Kenya) by 1.9 degrees C."

Global warming is expected to melt most of Africa's glaciers within the next few decades, which will reduce the already critically low amount of water available for farming. Long periods of drought followed by deluge rainfall have had devastating impacts in places such as Mozambique. Coastal areas in East Africa have suffered damage from storm surges and rising sea levels. The World Bank projects that by 2050 86 million people could be displaced by climate-related changes.

WARNING

Unfortunately, because of pervasive poverty and the historic scourge of HIV/AIDS and now of COVID, many areas of Africa lack the necessary resources to help people living there cope with climate change. And the effects of global warming may act as a barrier to development and aggravate existing problems. At present, as many as 400 million (or 33 percent of the continent's population) lack drinkable water, according to the World Resources Institute. The IPCC projects that some countries could see a 50-percent drop in crop yields over the same period and a 90-percent drop in revenue from farming by the year 2100. (We look at how developing nations are affected by and are addressing global warming in Chapter 12.)

Asia

More people call Asia home than any other continent — 4.7 billion in all. This high population, combined with the fact that most of Asia's countries are developing, means that a lot of people won't be able to sufficiently adapt to climate change impacts. As in Africa, climate change will bring pressures to the continent that will slow down development.

Here are some impending concerns for many parts of the continent:

>> **Future availability of drinkable water:** This has been and continues to be a major problem because of population growth, pollution, and low or no sanitation. The IPCC projects that anywhere from 120 million to 1.2 billion people may find themselves without enough drinkable water within the next 42 years, depending on the severity of climate change. Already, rising temperatures are causing glaciers in the Himalayas to melt. These disappearing glaciers, which are the water supply to 2 billion people, are also contributing to increased avalanches and flooding.

>> **Rising sea levels for coastal Asia:** The IPCC reports that mangroves, coral reefs, and wetlands will be harmed by higher sea levels and warming water temperatures. Unfortunately, this slightly salty water won't be good for freshwater organisms, as a whole. (See Chapter 8 for more about the impact global warming will have on the oceans.)

>> **Illnesses:** They're also expected to rise because of global warming. Warmer seawater temperatures could also mean more, and more intense, cases of cholera. Scientists project that people in South and Southeast Asia will experience more cases of diarrheal disease, which can be fatal. (Chapter 9 offers more information about how global warming might increase the environmental conditions that promote the spread of diseases.)

Australia and New Zealand

If you ask an Australian or a Kiwi about global warming, you probably won't get any argument about its negative effects. According to the World Wildlife Fund (WWF), Australia has experienced increased extreme and deadly bush fires, heat waves, less snow, and changes in rainfall. Extreme drought conditions persisted from 2003 to 2012 and from 2017 to today. This heat and lack of precipitation will likely worsen while global warming's effects intensify.

REMEMBER

The ozone layer in the Earth's atmosphere is sort of like sunscreen for the planet — *ozone* intercepts some of the ultraviolet radiation that causes sunburn and skin cancer. The use of chlorofluorocarbons for refrigerants and other purposes caused the ozone layer to get thinner, resulting in an ozone hole over Australia and New Zealand. Partly as a result, Australians have the highest incidence of skin cancer on Earth. In 1987, the nations of the world came together to regulate the use of these chemicals, and their concentration in the atmosphere continues to decrease and the ozone layer is making a comeback. But now, increasing average temperatures in Australia and New Zealand are compounding these effects — one problem reduced by international cooperation is still affected by the lack of international cooperation on another.

Climate change has also strongly affected the ocean. Sea levels have already risen 2.8 inches (70 millimeters) in Australia since the 1950s, and increasing ocean temperatures threaten the Great Barrier Reef. The reef is at risk of bleaching, half its coral has disappeared since 1995 and the possibility that it may be lost altogether is becoming more real. (See Chapter 8 for details.)

Small islands

You probably aren't surprised to hear that when it comes to climate change, rising sea levels and more extreme storms create an enormous risk for small islands everywhere, such as the South Pacific island of Tuvalu. Some islands will simply disappear due to rising sea levels if global efforts to limit global warming aren't successful. Here are other climate-related concerns for small island nations:

>> **Forests vulnerable to major storms:** Storms can easily topple island forests because a forest's small area doesn't provide much of a buffer and the root systems of trees are generally quite shallow on islands.

>> **Limited resources:** Some islands can't adapt physically and/or financially.

>> **Proximity of population to the ocean:** At least 50 percent of island populations live within a mile (1.5 kilometers) of water, and these populations are threatened by rising sea levels. Tsunamis (they used to be called *tidal waves*) caused by earthquakes and volcanoes, and storm surges from hurricanes and typhoons do much more damage when the ordinary level of the sea surrounding an island is even a little higher than it used to be.

>> **Risks to drinkable water:** The intrusion of ocean saltwater because of rising sea levels could contaminate islands' drinkable water, which is already limited on most islands.

>> **Reliance on tourism:** Beach erosion and coral reef damage, two possible effects of climate change, would undermine tourism, which many islands rely on for their source of income.

>> **Vulnerable agriculture:** Island agriculture, often a key part of the local economy, is extremely susceptible to harmful saltwater intrusions, as well as floods and droughts.

Polar regions

The planet's polar regions are feeling climate change's effects more intensely than anywhere else in the world. Warming temperatures are melting the ice and thawing the *permafrost* (the permanently frozen layer of earth in northern regions of Alaska, Canada, and Siberia) that used to be solid ground.

The Arctic is home to many changes brought on by global warming, including the following:

>> **Lost traditions:** Some indigenous people who make their homes in the Arctic are having to abandon their traditional ways of life. The Arctic ice and ecosystem are both core to many of these people's cultures and livelihoods. For more on this issue, flip ahead to Chapter 9.

>> **Melting ice:** The Greenland ice sheet is melting, adding to sea level rise. Arctic ice is also steadily losing ice volume. All of this melting is diluting ocean waters and affecting ocean currents.

>> **New plant life:** Greenery and new plants have been appearing in the Arctic in recent years. The *tree line* (where tree growth use to end and tundra began) is shifting farther north, but the soil isn't there to support a forest. Soils and ecosystems take thousands of years to develop — the changes happening now are rapid and unpredictable.

GOOD
NEWS

Some people look forward to the changes that the Arctic is experiencing. Now that so much sea ice has melted, ships can navigate the Arctic Ocean more efficiently, taking shorter routes. Without any sense of irony, oil companies now keenly anticipate being able to reach more fossil fuels below what used to be unreachable areas because of ice cover. Communities in the Arctic may be able to harness river flows that have been boosted or created by ice melt to run hydroelectric power. But these short-term economic developments can't outweigh the negative planetary impacts.

In the Antarctic, some scientists project major change because of global warming, thinking there's a chance that the western Antarctic ice sheet might collapse by the end of the 21st century. The western Antarctic ice sheet is simply enormous. It contains about 768,000 cubic miles (3.2 million cubic kilometers) of ice, about 10 percent of the world's total ice. It appears to be weakening because warmer water is eroding its base. For the first time in the 2021 Sixth Assessment Report, IPCC scientists accepted as plausible, but not likely, that the entire sheet could melt. The Greenland ice sheet is also melting — quickly. Both the western Antarctic and Greenland ice sheets are adding to sea level rise.

The melting polar ice is also endangering many species, such as polar bears and penguins, which rely on the ice as a hunting ground. (Chapter 8 offers more information about the ways the polar animals are being affected by global warming.)

Positive Politics: Governments and Global Warming

Governments are often the first institutions that the public looks to for big solutions. Governments represent the people of a region, after all, and are expected to make decisions for the good of the public. So, governments need to be able to respond to global warming effectively. Climate change is a very big problem for which no one has all the answers. Despite this challenge, governments around the world are willing to play their part — and it's an important one.

Governments need to take the lead. The next sections lay out some of the necessary actions at all levels from your local water authority to the international institutions.

Making a difference from city hall to the nation's capital

All levels of government, from cities and towns, to states and provinces, to countries, have the ability to affect taxes and laws that can help in the fight against climate change:

>> **Local governments:** Can implement and enforce city building codes, improve public transit systems, and implement full garbage, recycling, and composting programs.

>> **Regional governments:** Can set fuel efficiency standards, establish taxes on carbon dioxide emissions, and set efficient building codes.

>> **Federal governments:** Can lead on the largest of issues, such as subsidizing renewable energy sources, removing subsidies from fossil-fuel energy sources, taxing carbon, and developing national programs for individuals who want to build low-emission housing. Federal governments can also set standards and mandatory targets for GHG reductions for industry, provinces, and states to follow.

The most effective governments work with each other — partnerships between cities, states, and countries exist around the world, supporting one another while they work on the same projects. To read more about what governments can do and are already doing, check out Chapter 10.

Working with a global government

Countries must work together through global agreements to deal with, and conquer, a problem as urgent, complex, and wide-sweeping as climate change. Global agreements create a common level of understanding and allow countries to create collaborative goals, share resources, and work with each other towards global warming solutions. No one country can solve climate change on its own, just like no one country created global warming in the first place.

The core international law around climate change is the 1992 UN Framework Convention on Climate Change (UNFCCC) and a series of subsequent agreements, from the 1997 Kyoto Protocol to the current 2015 Paris Agreement. Countries have agreed that globally they will hold to as far below 2 degrees C as possible and preferably to no more than 1.5 degrees. But, collectively, despite marked progress in some nations, particularly within the European Union, the world's countries aren't on track to deliver on these goals.

The international discussions are ongoing; government representatives meet on an annual basis for the United Nations Climate Change Conference. These targets we re-affirmed at the last such meeting in Glasgow, Scotland in 2021. We discuss just what goes on at those meetings in Chapter 11.

Helping developing countries

The effects of climate change are taking a particularly heavy toll on the populations of *developing countries* — countries with little or no industry development and a weak or unstable economy. These countries, which are primarily located in Latin America, Asia, and Africa, have fewer financial resources to recover from events such as flooding, major storm damage, and crop failures. Money that these nations have to spend paying for the effects of global warming is money that they can't spend building their economies.

Developing countries have little or no major industry development, for the most part (although China has overtaken the United States as the world's largest polluter), so they don't add many GHG emissions to the atmosphere. Even China, with its growing industry, lags far behind the emissions of industrialized nations on a per-person basis. Because industrialized countries have been the primary GHG emitters, they have the main responsibility for reducing emissions, and they can also play a role in helping developing countries shift to renewable energy sources and adapt to climate impacts. For more about how developing nations are addressing climate change, see Chapter 12.

Solving the Problem

Everyone can play a part in slowing down global warming, and humanity doesn't have time to start small. Solving climate change requires a major commitment from everyone — from big business and industry to everyday people. Combined, the following changes can make the necessary difference.

Changing to alternative energies

Fossil fuels (see Chapter 4) are the primary source of the human-produced GHGs causing global warming. Although they've fueled more than a century of human progress, it's time to leave them with the dinosaurs. Fortunately, a wide array of energies is waiting to take the place of oil, coal, and gas.

REMEMBER

Here's a list of *renewable resources* — energy that doesn't run out, unlike fossil fuels, and doesn't pump more carbon into the atmosphere:

>> **Geothermal:** Jules Verne was wrong; the center of the planet doesn't contain another world, but it does have plenty of heat. People can use that heat to boil water to produce steam that propels turbines and generates electricity. Even areas without geo-heat sources to boil water can heat homes through *geothermal* energy (the warmth of the earth).

>> **Hydro:** People can harness *hydropower,* or water power, to turn turbines and create electricity.

>> **Solar:** Humanity can use the sun's warmth in a few ways. Solar cells, like you see on some roofs, can convert sunlight to electricity. People can also heat buildings and water with the sun's direct heat.

>> **Waste:** Garbage is more than just trash. It offers astounding possibilities. People can harness the methane emitted from dumps, burn the byproducts of agriculture as fuel, and even use old frying oil as a type of diesel.

>> **Wind:** Remember that pinwheel you had as a kid? Giant versions of those wheels are popping up all over the world as wind turbines, generating clean electricity for homes, businesses, and entire energy grids.

Feeling charged up? Check out Chapter 13 to further explore the renewable-energy possibilities, and Chapter 17 to see how rapidly some of these changes are happening already.

Getting down to business

Big industry is the largest contributor to GHG emissions, and it can make the biggest contributions to the fight against global warming. Although some of the changes that businesses can make may have an initial impact on the businesses' pocketbooks, many of these changes may even save businesses money in the long run.

Industrial-strength solutions

The greatest immediate change businesses and industries can make is to improve their efficiency. Companies waste a lot of energy powering antiquated equipment, heating poorly insulated buildings, and throwing out materials that they could recycle. Chapter 14 details some of the ways that companies can pull up their socks and make smarter use of energy, and it also shares some impressive success stories.

Ideally, renewable energy will ultimately power industry. Industry dependence on fossil fuels must rapidly decrease. Currently, some scientists and industries are trying to store carbon emissions underground. This solution is controversial, however. We consider the issue in Chapter 13.

Green fixes for forestry and farming

The forestry and agriculture industries can do more than just cut back on their GHG emissions; they can actually increase the amount of carbon that's absorbed from the planet's atmosphere. (See Chapter 2 to take a ride on the carbon cycle and understand the critical role that plants play in keeping Earth livable.)

REMEMBER

Around the world, forests are being cut down, removing valuable *carbon sinks,* which absorb carbon from the atmosphere. Where they harvest trees, logging companies need to explore methods other than clear-cutting; selectively harvesting trees enables forests to continue to thrive. In other countries, particularly in South America, people are clearing forests for farmland. Losing those forests is particularly costly for the atmosphere because, unlike forests in more temperate climates, these rainforests absorb carbon year-round. Deforestation methods have to change.

Farming's solution for global warming is dirty — or how dirt is treated. Believe it or not, a simple action like excessively tilling the land causes carbon to be released into the atmosphere. And when farmers add GHG-laden fertilizers to the soil, they release even more emissions. By cutting back on tilling the land and using less fertilizer, farmers can be a potent part of the solution to climate change. *Regenerative agriculture* (an approach to farming that works to rebuild topsoil) can play a big part in avoiding climate disaster.

Making it personal

You're a vital part of the climate change solution, too. As a citizen, you can ensure that governments recognize the importance of global warming and follow through on their promises. As a consumer, you can support companies that are making the biggest strides in fighting climate change and encourage other companies to make reducing GHGs a priority. If you're really passionate about having your voice heard, you might even want to consider joining a group dedicated to spreading the word about global warming. We tell you how you can get involved in Chapter 15.

TIP

You can also make many changes in your daily life — some that seem small, some less so — that cut back on the carbon emissions for which you're responsible. You're probably already familiar with many of the little steps you can take to be more climate friendly:

>> **Making your home more energy efficient:** Better insulate your roof, basement, and walls; seal your windows; and replace your old light bulbs with LEDs.

>> **Reducing the amount of garbage you produce:** Take a reusable bag with you when you shop, buy unpackaged goods, and recycle and reuse materials.

>> **Using energy wisely:** Turn off lights and appliances when you're not using them, use the air conditioner less in the summer, and turn down the heat in the winter.

Did you know that many of your appliances are gobbling electricity, causing the emission of GHGs, even when those appliances are turned off? Or that putting a lid on pots on your stove makes your food cook more efficiently?.

Not every action that you can take to cut back on your GHG emissions is manageable — not everyone can yet buy a hybrid or electric car or build a home that doesn't rely on major power producers for energy. But, hopefully, we suggest some options in this book that fit your situation and can help you to make a difference.

Global warming affects everyone, and everyone can play an important role in stopping it. Balance the doom and gloom and — start thinking about the exciting opportunities you have to make a change.

Chapter **2**

Looking Closely at Greenhouse Earth

What you've read or seen may make you think that the greenhouse effect and greenhouse gases (GHGs) are all bad. Actually, GHGs have long been the good guys.

Planet Earth is a tiny warm dot in vast frigid space. The atmosphere keeps the Earth warm because the atmospheric gases trap the heat of the sun, just as greenhouse glass does. Inside greenhouse Earth life is comfortable. Anywhere without a nice balance of GHGs, like say Venus (hot, hot, hot) or Mars (really cold), not so much.

When it comes to the survival of all living things on Earth including us humans, the home planet needs to be like the porridge in *Goldilocks and the Three Bears:* not too hot or too cold, but just right.

GHGs become a problem only when the atmosphere contains too much of them, which is happening today. Industries and farms, cities, and garbage dumps are pumping out an array of gases — carbon dioxide, methane, nitrous oxide, and a host of other substances. Humanity has knocked off kilter the life-preserving cycle that makes sure the Earth's atmosphere has just enough carbon dioxide, the star GHG.

This chapter is the A–B–Cs of climate science. This is basic science, and it's less complicated than you think! You know (from third grade) how plants use energy from the sun to soak up carbon dioxide and release oxygen. You probably remember the big word — photosynthesis. This chapter gives you the basics.

Examining Greenhouse Effect 101

If you want to understand the greenhouse effect, the best place to start is with the object that provided this analogy in the first place, the greenhouse. A greenhouse works by letting in sunlight, which plants and soil absorb, thus heating up the greenhouse. The panes of glass ensure that the warmer air doesn't escape the greenhouse, or does so very slowly. If you've ever parked your car with the windows rolled up on a sunny day, you've experienced this effect. When you open your car door, you're hit with a blast of hot air. The windows of your car have acted like the panes in a greenhouse, letting sunlight in, which heats the car, and then trapping that heat.

Certain gases in the atmosphere trap the sun's heat in a similar way. These particular gases are called *greenhouse gases* because they cause this greenhouse effect. The Earth is bombarded by radiation from the sun. Some of this radiation can be seen (think visible light), and some of it can't be (ultraviolet light, for example).

TECHNICAL STUFF

Very hot bodies give off different amounts of energy than cold ones do. A basic law of physics says that everything gives off radiative (mostly heat or light) energy, and how much energy it emits depends on its temperature. The sun, for example, is a toasty 10,300 degrees Fahrenheit (5,700 degrees Celsius) — a little bit hotter than people are used to here on Earth. So, the sun gives off a lot of radiative energy, and the Earth gives off very little. Earth is warm mostly because of the heat it gets from the sun — most of the sun's radiative energy actually zooms right through the atmosphere to the Earth's surface. (The helpful high level ozone layer protects us by absorbing a lot of the harmful ultraviolet rays.)

A portion of this radiation, about 30 percent on average, bounces off clouds, ice, snow, deserts, and other bright surfaces, which reflect the sun's rays back into outer space. The other 70 percent is absorbed by land or water, which then heats up. And the Earth emits some of that heat — in the form of infrared radiation (electromagnetic waves most commonly known as heat). The unique qualities of the GHGs come into play: The GHGs absorb some of the escaping infrared radiation in the lower atmosphere, and re-radiate part of that back down. So, less of the radiation from the Earth's surface gets to outer space than it would have without those gases, and that energy remains in the atmosphere and returns to the Earth's surface — making both the atmosphere and Earth itself warmer than they would be otherwise.

To see the greenhouse effect in action, look at Figure 2-1.

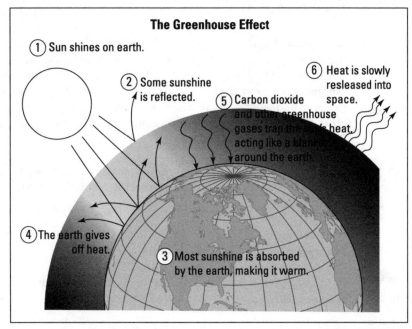

The Greenhouse Effect

(1) Sun shines on earth.

(2) Some sunshine is reflected.

(6) Heat is slowly resleased into space.

(5) Carbon dioxide and other greenhouse gases trap the sun's heat, acting like a blanket around the earth.

(4) The earth gives off heat.

(3) Most sunshine is absorbed by the earth, making it warm.

FIGURE 2-1:
The greenhouse effect in action.

© John Wiley & Sons, Inc.

If the planet had no atmosphere or GHGs, humanity would be left out in the cold. The Earth wouldn't be able to keep any of the heat that it gets from sun. Thanks to GHGs, humanity is kept reasonably warm, enjoying an average temperature of 59 degrees Fahrenheit (15 degrees Celsius), some 62.6 to 64.4 degrees Fahrenheit (17 to 18 degrees Celsius) warmer than without GHGs.

REMEMBER

This natural greenhouse effect and the ozone layer allow life to exist on Earth. Without the greenhouse effect, the Earth would be too cold. And without the ozone layer, life couldn't survive the sun's ultraviolet radiation.

Too much GHG turns the heat up beyond that to which societies and ecosystems have become adjusted. The atmosphere on the planet Venus is 96 percent carbon dioxide (the key GHG that we talk about in the following section). Because of Venus's concentration of GHGs and its relatively close proximity to the sun, it's extremely hot — surface temperatures of up to 500 degrees Celsius. Meanwhile, the atmosphere on Mars has 95 percent carbon dioxide, but it's very thin, and the planet's position is farther away from the sun than Earth, so it's extremely cold — a chilly −80 degrees Fahrenheit (−60 degrees Celsius).

Focusing On Carbon Dioxide: Leader of the Pack

Earth's atmosphere contains 24 different GHGs, but just one of them accounts for the overwhelming majority of the effect: carbon dioxide (or CO_2, for short). This gas accounts for about 63 percent of the GHG warming effect in the long run. (In the short term, over the last 5 years, it has accounted for 91 percent. See Table 2-1 for more on the intensity of gases over time.) If you're itching to know about the other 37 percent of greenhouse warming, check out the section "Checking Out the Other GHGs," later in this chapter.

TECHNICAL STUFF

Water vapor, not carbon dioxide, is technically the GHG with the biggest impact. But human activities don't directly affect in a significant way water vapor in the atmosphere.

Given the important role that carbon dioxide plays in warming the Earth, you may be surprised by how little of it is in the atmosphere.

TECHNICAL STUFF

In fact, 99.95 percent of the air that humans breathe (not including water vapor) is made up of

>> **Nitrogen:** 78 percent

>> **Oxygen:** 21 percent

>> **Argon:** 0.95 percent

Carbon dioxide, by contrast, currently makes up only 0.0412 percent of all the air in the atmosphere. Human activities have helped increase that concentration from pre-industrial times, when it was about 0.0280 percent.

When scientists talk about air quality and the chemistry of the atmosphere, they often use the term parts per million (ppm). So, currently out of every million parts of air, only 412 are carbon dioxide. That's not much carbon dioxide, but what a difference it makes! Until recent changes in atmospheric chemistry caused by human activity, for around the last million years, carbon dioxide concentrations never exceeded 285 ppm. It's like the hot pepper you put into a pot of chili — just right is just right, but if you have just a little too much, watch out.

The next sections explain the carbon cycle. If not for humans digging up the fossilized stored carbon of millennia ago — called fossil fuels — the Earth's carbon cycle would have remained in balance. However, now the carbon cycle is out of balance. And that's a key reason the weather and climate are no longer so hospitable.

Looking at the carbon cycle

Carbon dioxide occurs naturally — in fact, you, and every animal and insect on Earth that breathes in air, produce carbon dioxide every time you exhale. You inhale oxygen (and other gases), which your body uses as a nutrient, and you breathe out what your body doesn't need, including carbon dioxide. You aren't alone in using this process. But other organisms, mostly plants, suck carbon dioxide out of the air. Trees and grasses, for example, take in carbon dioxide and give out oxygen — the complete opposite of what people do.

The *carbon cycle* is the natural system that, ideally, creates a balance between carbon emitters (such as humans) and carbon absorbers (such as trees), so the atmosphere doesn't contain an increasing amount of carbon dioxide. It's a huge process that involves oceans, land, and air. Life as humans know it — from microscopic bugs in the oceans to you and me, and every fern and plant in between — would disappear without this cycle. You can think of the carbon cycle almost as the Earth breathing in and out.

REMEMBER

The carbon cycle is called *in balance* when roughly the same amount of carbon that's being pumped into the air is being sucked out by something else. The atmospheric concentration of carbon dioxide was historically at a concentration of 280 ppm — carbon dioxide concentrations have fluctuated up and down through natural processes, but 280 ppm has been about the highest recorded concentration for the past 800,000 years — until recently when humans started to increase the concentration. (We look at how humans contribute carbon dioxide to the atmosphere in Part 2.)

A RECIPE THAT GIVES YOU GAS

Carbon dioxide is composed of one carbon atom and two oxygen atoms. Present in the atmosphere as an odorless, colorless gas, it can also exist in solid form (think dry ice) and, when kept under pressure, in liquid form (the bubbles you see in champagne or a can of soda are carbon dioxide escaping after you uncork the bottle or open the can and remove the pressure).

When trees take up carbon through photosynthesis, they're called *carbon sinks.* Plants aren't the only carbon sinks, however. Figure 2-2 shows how the ocean, plants, and soil all act as carbon sinks, removing carbon from the atmosphere. They also store, or *sequester,* carbon, and they store the carbon in a carbon *reservoir.* For example, the ocean holds about 38,000 billion metric tons of carbon in its reservoir.

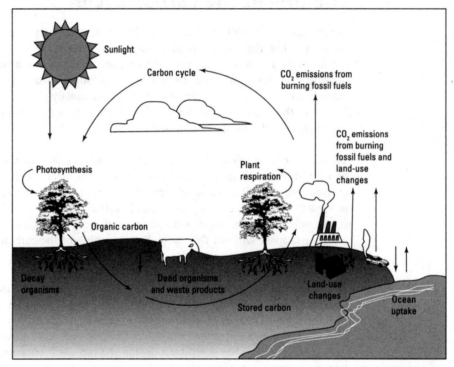

FIGURE 2-2:
The carbon cycle.

© *John Wiley & Sons, Inc.*

The next sections explain how everything is connected. What goes on in the atmosphere isn't isolated from everything else in the world. It's like that old spiritual: "Dem bones, dem bones, dem dry bones. Toe bone connected to the foot bone. Foot bone connected to the heel bone. Heel bone connected to the ankle bone. Ankle bone connected to the leg bone . . ." and on and on. Earth is like that too. Everything is connected. The ocean is connected to the atmosphere. The plant life — forests on land and green things under water — are all connected. Changes in the atmosphere have impacts on the oceans, the forests, the clouds, and the soil. And vice versa.

Under the deep blue sea

The ocean is the biggest carbon sink on Earth. So far, it has tucked away about 90 percent of all the carbon dioxide in the world. If that gas was in the atmosphere, not underwater, the world would be a lot hotter.

The exchange of carbon dioxide between the ocean and the air happens at the surface of the water. When air mixes with the surface of the ocean, the ocean absorbs carbon dioxide because carbon dioxide is *soluble* in water (that is, carbon dioxide can be absorbed by water). And, in fact, the seas' ability to absorb carbon dioxide is referred to as the *solubility pump* because it functions like a pump, drawing carbon dioxide out of the air and storing it in the ocean.

The ocean also acts as a biological pump to remove carbon dioxide from the atmosphere. Plants close to the surface of the ocean take in carbon dioxide from the air and give off oxygen, just like plants on land. (We discuss this process, known as photosynthesis, and the role that plants play in the carbon cycle in the following section.) Phytoplankton are microscopic plants that live in water. You may know them as algae, most commonly seen as the greenish clumpy plants that float around on ponds and other water. Phytoplankton have short but useful lives. If other organisms don't eat them, they simply die within just a few days. They then sink to the ocean floor, mix into the sediment, and decay. The carbon dioxide that these plants absorb during their brief lives is well and truly sequestered after their little plant bodies are buried.

Each year, the oceans put away about another 2 billion metric tons of carbon dioxide. Figure 2-3 demonstrates how the ocean interacts within the carbon cycle. According to the World Economic Forum, recent research suggests that the 2 billion metric tons of carbon a year may actually be an underestimate. It could be as much as .9 billion metric tons more — every single year. Sometimes it's expressed as ten Hiroshima-size nuclear bombs worth of energy in heating — absorbed by the oceans per second. Scientists writing about the huge amount of energy stored by the oceans search of explanations in equivalents that people can grasp — like nuclear bombs. That's because the science is expressed in unfamiliar terms. Describing the energy in terms of nuclear bombs is a scientist's way of expressing that the ocean absorbed 20 sextillion joules of heat in 2020.

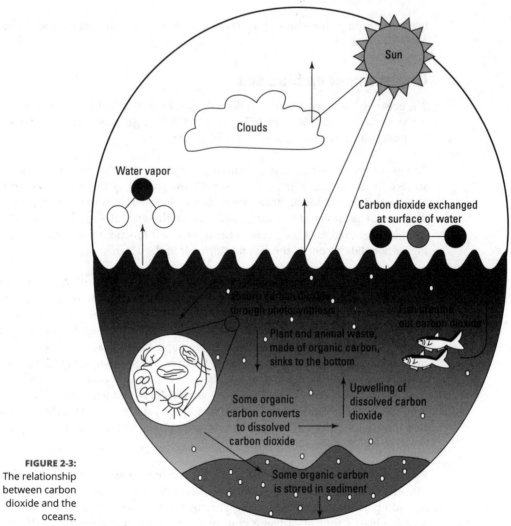

Clouds

Sun

Water vapor

Carbon dioxide exchanged
at surface of water

absorb carbon dioxide
through photosynthesis

Fish breathe
out carbon dioxide

Plant and animal waste,
made of organic carbon,
sinks to the bottom

Some organic
carbon converts
to dissolved
carbon dioxide

Upwelling of
dissolved carbon
dioxide

Some organic carbon
is stored in sediment

FIGURE 2-3:
The relationship
between carbon
dioxide and the
oceans.

Why people couldn't survive without plants

You may not have realized back in elementary school that when you were reading
about photosynthesis, you were actually getting the basics of modern climate science. (*Photosynthesis* occurs when plants take in energy from the sun and carbon
dioxide from the atmosphere and turn it into oxygen and sugars.) Figure 2-4 may
jog your memory.

Trees are the planet's biggest and most widespread plants, and the forests are
wonderful carbon sinks.

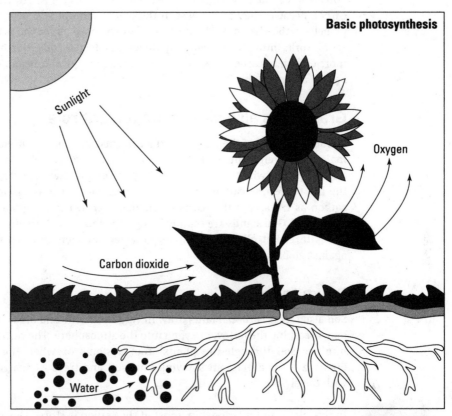

Basic photosynthesis

Sunlight

Oxygen

Carbon dioxide

Water

FIGURE 2-4:
The process of
photosynthesis.

© *John Wiley & Sons, Inc.*

REMEMBER

The most effective carbon-trapping forests are tropical, such as those in Brazil and other South American countries. Most tropical forests are called rainforests (although not all rainforests are tropical). *Rainforests* grow in regions that get more than 70.9 inches of rain each year. Because of all the rain they get, these dense, rich forests are full of biodiversity. And because of the tropical climate, which is always warm, these tropical forests work year-round. The tireless work that these trees do to sequester carbon is just one of the reasons to protect the tropical rainforests.

Mangrove forests are another little appreciated forest ecosystem. They're also tropical, but they're rooted in water. Research shows they may actually be four times more effective in sucking up carbon than tropical forests on land. But they're at risk. About a third of the world's mangroves have been removed — mostly for tourism developments to create beaches and for farming shrimp in toxic shrimp ponds. Planting mangroves helps nature, creates homes for fish, protects coastal communities from big storms, and fights climate change. And unlike the forests on land, they can't burn up because they live in water.

Forests in Canada, the United States, and Russia aren't as effective at soaking up carbon because they take a rest in the winter but are still very important in the planet's carbon balance. The northern forests make up for the reality of their seasonal work, through the relatively richer and deeper soils. Northern forests store more carbon in carbon reservoirs, even though tropical forests take up more carbon on an annual basis.

Grasslands also play an important role

About 40 percent of the Earth's surface is grassland, mostly used for grazing animals. They aren't only cattle — sheep, goats, yaks, camels, llamas and alpacas, and the people who tend to them all depend on grasslands. Grasslands are one of the most effective carbon sinks — the deep roots of grasses can store as much carbon as trees, and that carbon remains even when the grasses are grazed or burned off. Grasslands are rapidly being depleted by conversion to more intensive agriculture. But climate scientists now see their enormous value in the fight against global warming.

Down to earth

Soil also stores carbon. Plants draw in carbon dioxide and break it down into carbon, breathing the leftover oxygen into the atmosphere. The carbon makes its way into the soil through the plants' root systems or when the plant dies. See Figure 2-5 for a diagram showing how soil and trees exchange carbon dioxide with the air.

REMEMBER

In this plant-soil relationship, most of the carbon is stored close to the top of the soil. Tilling the soil (mixing it up) exposes the carbon in the ground to the oxygen in the air, and these two elements immediately join to form carbon dioxide.

Altogether, vegetation and soil store about a billion metric tons of carbon every year, and another 1.6 billion metric tons move in and out between the land and the air. So far, the plants, animals and soil have packed away 3.3 billion metric tons.

Investigating humanity's impact on the carbon cycle

A lot of the carbon dioxide in the atmosphere is natural (you're breathing some out, right now), but human activities also contribute plenty of the gas (we discuss these activities in Part 2). Historically, the carbon dioxide that people put into the air was pretty much soaked up by the carbon sinks, and the amount of carbon dioxide that was around before people started building factories had been fairly steady since the beginning of human civilization.

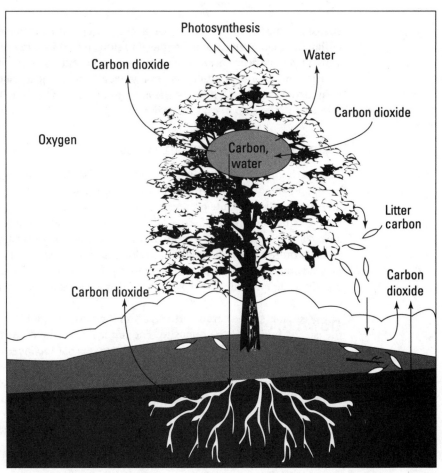

Photosynthesis

Carbon dioxide

Water

Carbon dioxide

Oxygen

Carbon, water

Litter carbon

Carbon dioxide

Carbon dioxide

FIGURE 2-5:
How trees and soil work side-by-side with carbon dioxide.

© *John Wiley & Sons, Inc.*

Producing industrial amounts of carbon dioxide

Since the Industrial Revolution went into full swing around 1850, the amount of GHGs in the atmosphere has risen drastically. Due to burning fossil fuels, as well as clearing forests, people have almost doubled the carbon dioxide emissions in just over a century, and today, carbon dioxide levels are higher than they have ever been in recorded history (see Chapter 4 for more about fossil fuels). In fact, atmospheric carbon dioxide levels are higher today — a 45 percent increase — than at any time in the past 800,000 years. (Carbon dioxide levels were much higher millions of years ago, however. We talk about the history of carbon dioxide levels in greater detail in Chapter 3.)

Carbon dioxide concentration levels are currently at about 412 ppm, and they rose at an average of 2 ppm per year between 2000 and 2021 because of increasing emissions due to human actions. On average, in the 1980s, globally, people put 7.2 billion metric tons of carbon dioxide emissions into the air every year — and those emissions have been increasing every year — now at 43 billion, per the Center for Climate and Energy Solutions. So many people are using so much energy, mostly in industrialized countries, that the amount of carbon that is being put into the air is knocking the carbon cycle off balance.

Plugging up the carbon sinks

The Earth's carbon sinks, which used to be able to handle everything oxygen-breathing creatures could throw at them, aren't able to keep up with humanity's increased carbon dioxide production. Studies presented through the Intergovernmental Panel on Climate Change (IPCC) reports suggest a bunch of different possible consequences, ranging from a theory that new plants might appear that can soak up more carbon dioxide to the idea that carbon sinks may become full and may no longer be able to absorb any more carbon dioxide. Like anyone who works overtime, carbon sinks could become weaker as they soak up more carbon dioxide.

The ocean has stored carbon effectively in the past, but global warming is causing the oceans to do just the opposite. The top layers of the oceans — the top 2,300 feet (700 meters) have warmed a lot since 1900. That top layer is now 1.5 degrees Fahrenheit (0.83 degrees Celsius) warmer. Carbon dioxide is less soluble in warm water. The oceans push the carbon dioxide that they can't dissolve into the air, instead. Data collected during the 1980s and 1990s suggested that both land and ocean sinks seemed to have kept up with growing emissions. However, more recent studies show that the carbon dioxide intake of some sinks, such as trees, is slowing down.

In addition to the warming impact, as carbon dioxide mixes in the top layers of ocean water, the oceans are getting more acidic. Carbon dioxide mixing with ocean water makes a chemical change to carbonic acid. Carbon dioxide is less soluble in warm water, so the acidification of oceans is worse in the colder regions, and this trend is super worrying. The increasingly acidic ocean makes it harder for sea creatures that live in shells to form those shells. Where Elizabeth and John live on Vancouver Island, aquaculture operations growing oysters and scallops have had to move the early stages of growing shells to the warmer waters of Hawaii, to then transport the scallops and oysters back to Vancouver Island's colder waters to reach maturity. This increased acidity is measurable. In 2021, the oceans are 25 percent more acidic than in 1900.

Sinks normally absorb about half of human-caused emissions. So, if these sinks were to weaken, or even stop absorbing, they'd leave a lot more carbon dioxide in the atmosphere, on top of our already-increasing emissions.

Checking Out the Other GHGs

Carbon dioxide may get all the press, but 23 other GHGs (in five main groups) also heat things up. Although they're present in much smaller amounts, these gases are actually far more potent, molecule for molecule, in terms of greenhouse effect. You might think of them as carbon dioxide on steroids. Table 2-1 shows you the power of some of these gases compared to carbon dioxide as the reference starting point with a global warming potential of 1.

TABLE 2-1 Global Warming Potential of GHGs

GHG	Global Warming Potential Over Time	
	20 years	100 years
Carbon dioxide (CO_2)	1	1
Methane (CH_4)	56	21
Nitrous oxide (N_2O)	280	310
Hydrofluorocarbons (HFC) Group of 13 gases	3,327	2,531
Perfluorocarbons (PFC) Group of 7 gases	5,186	7,614
Sulfur Hexafluoride (SF_6)	16,300	23,900

Source: United Nations Framework Convention on Climate Change, GHG Data, Global Warming Potentials, http://unfccc.int/ghg_data/items/3825.php

Because so many different types of GHG exist, people usually either talk about only carbon dioxide (because so much more of it exists than the others) or GHGs in terms of carbon dioxide equivalents — how small an amount of the gas you'd have to put into the atmosphere to have the same warming impact as the current level of carbon dioxide. Referring to all GHGs with this measurement makes assessing and measuring them that much easier. So, when we say "greenhouse gas" in this book, you can actually think of it as carbon dioxide equivalent emissions. No calculator needed.

TECHNICAL
STUFF

Measuring in carbon equivalents means, for example, that 1 unit of methane equals 21 units of carbon. In other words, 1 metric ton of methane is just as bad as 21 metric tons of carbon dioxide. Thus, methane is 21 carbon dioxide equivalents, or 21 metric tons of carbon dioxide.

These sections focus on the other GHGs that also heat up the planet. Water vapor has a big impact on warming and is taken into account by climate scientists. But it's the increase in carbon dioxide and methane that are pushing the global carbon balance into imbalance and to dangerous shifts that drive major climatic changes.

Methane (CH_4)

Methane (CH_4) accounts for four to nine GHGs of the overall 24 GHGs, according to the World Meteorological Organization, but it's 21 to 56 times more potent than carbon dioxide. Methane is to carbon dioxide what an espresso shot is to herbal tea. (See Table 2-1.)

Methane naturally occurs when organic materials, such as plant and animal wastes, break down in an *anaerobic* environment (an environment that contains no oxygen and includes the right mix of microbes and temperature). This breakdown creates methane, along with small amounts of other gases. The stomach of a cow, a landfill site, and a marsh are all prime examples of methane-producing environments.

How methane gets into the atmosphere

Two-thirds of all the human-made methane comes from agriculture, and about half of that amount comes from rice crops. If you've ever seen rice being grown, you may remember that it's planted in a flooded field. Any dead organic matter falls to the bottom of the paddy, which is a perfect airless environment for creating methane.

All cows, pigs, chickens, and other farm animals account for the rest. The food breaking down in their stomachs produces methane that they, shall we say, emit into the air — one way or another. All animals emit methane — yes, even you — but livestock's methane causes a problem because there are so many of these fairly big animals.

Humans also add methane to the atmosphere through treating wastewater and from landfills — all that garbage spews methane into the air while it breaks down.

People also use methane as a fuel. Natural gas is 90 to 95 percent methane, and when natural gas is extracted from the ground, some methane escapes into the air.

Methane stabilized for a while — still it didn't

Methane levels in the atmosphere had stabilized at 1.8 parts per million, but since the first printing of this book, record growth in methane has been experienced globally. In 2021, smashing previous records methane grew to 1.9 parts per million.

Much of this increase was due to the use of novel technology (referred to as *hydraulic fracturing* or *fracking*) to get gas out of bedrock formations like shale. The boom in natural gas fracking has accelerated the move away from coal by electric utilities in the United States and Canada. But it isn't clear win for the climate. (Read more about using natural gas as a fuel in Chapter 4.)

Fracking threatens water quality with chemical contamination, and research has shown that it has even caused in increase in earthquakes. Turning the Earth into a pincushion has some worrying side effects: The process of fracking bombards the bedrock formations with jets of water, containing special chemicals to get the gas out of the substrate and into pipelines headed to making liquified natural gas (LNG). The process of fracking uses so much energy that even though using LNG for energy at the end point is relatively clean, the so-called *upstream* (or process of extraction of LNG) has the same GHG impact as burning coal.

WARNING

A lot of methane is frozen into the ground of the Arctic, trapped by the *permafrost* (any ground that remains completely frozen for more than two years). Rising northern temperatures, however, are melting the soil. When the Arctic land thaws, it becomes swampland — and it starts dishing out methane like hotcakes. We look at this problem in greater detail in Chapter 7.

Nitrous oxide (N_2O)

The amount of nitrous oxide (N_2O) in the atmosphere is even smaller than the amount of methane, but it accounts for about 7 percent of the overall greenhouse effect. The greenhouse effect of nitrous oxide per unit is almost 300 times more potent than that of carbon dioxide. This gas is actually still going up at a rate of 1 part per billion (ppb) each year — as of 2018, it was at 331 parts per billion. Increasingly, nitrous oxide comes from human activities.

The following are ways nitrous oxide appear in the atmosphere:

>> In agriculture, farmers encourage those natural bacteria to produce more of the gas through soil cultivation and the use of natural and artificial nitrogen fertilizers. The biggest source of nitrous oxide, natural or human-made, is fertilizers used in agriculture. Fertilizers count for 60 percent of human-made sources and 40 percent of sources overall.

>> Ocean- and soil-dwelling bacteria produce nitrous oxide naturally as a waste product.

>> Dentists use nitrous oxide as an anesthetic. (Laughing gas is nitrous oxide — not so funny now, is it? Though it's in such small amounts that you don't need to worry about your root canal adding to global warming.) Industrial processes (to create nylon, for example) also produce nitrous oxide.

>> Humans add a lot of nitrous oxide to the atmosphere by using automobiles. Ironically, cars produce the gas as a side result of solving another environmental problem — see the nearby sidebar for the scoop.

Hexafluoro-what?

Hydrofluorocarbons. Perfluorocarbons. Sulfur hexafluoride. Try saying those names three times fast. They're as hard to say as they are effective at trapping heat. These three types of gases are all human-made and don't exist naturally in the atmosphere. They come from a number of different industrial processes that create air pollution.

Almost all car air-conditioning systems use the 13 hydrofluorocarbons (HFCs). (We look at how industry emits these GHGs in detail in Chapter 5.) Most of the seven perfluorocarbons (PFCs) are by-products of the aluminum industry. Sulfur hexafluoride (SF_6) comes from producing magnesium, and many types of industry use it in insulating major electrical equipment.

HOW GOOD INTENTIONS INCREASED NITROUS OXIDE EMISSIONS

Fossil fuels (which we discuss in greater detail in Chapter 4) contain nitrogen. When cars burn gasoline, they give off the nitrogen-based chemicals nitrogen monoxide (NO) and nitrogen dioxide (NO_2) — together known as NO_x gases. These NO_x gases create acid rain and smog in cities.

In response to these environmental problems, governments in North America forced car companies to put catalytic converters in all their cars. Catalytic converters convert smog-causing chemicals into other chemicals that aren't as damaging to our lungs and don't cause acid rain.

Unfortunately, these catalytic converters turn NO_x gases, which don't have an effect on climate change, into nitrous oxide (N_2O), which does! (One more reason, if you need it, why everyone should drive less.)

Other players on the GHG bench

The two GHGs that we talk about in the following sections do play a role in climate change, but they aren't on the United Nations list of 24 GHGs and get left aside in most discussions about the impact of GHGs on global warming — not for scientific reasons, but because of decisions made in international negotiations.

Water vapor

As we discuss in the section, "Focusing On Carbon Dioxide: Leader of the Pack," earlier in this chapter, water vapor is a huge player in the greenhouse effect. As shocking as it may seem, good ol' H_2O (two parts hydrogen, one part water) causes the majority — 60 percent — of the planet's greenhouse effect. But the ramped-up threat of climate change isn't tied to water vapor. Water vapor remains an essential reason the planet is warm enough to sustain current life forms.

Unlike the production of the other GHGs, humans don't directly cause the increase of water vapor. But the other gases that are produced heat up the atmosphere. When plants, soil, and water warm up, more water evaporates from their surfaces and ends up in the atmosphere as water vapor. A warmer atmosphere can absorb more moisture. The atmosphere will continue to absorb more moisture while temperatures continue to rise. See Figure 2-6.

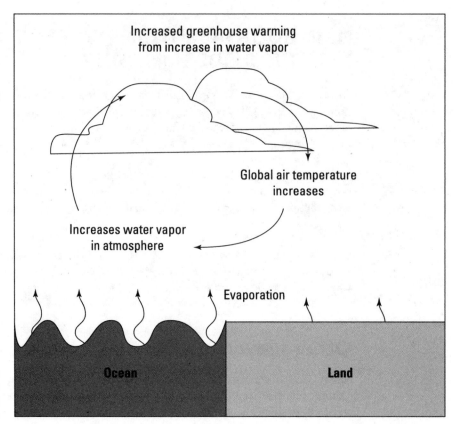

FIGURE 2-6:
Water evaporates
and lingers in the
atmosphere.

**TECHNICAL
STUFF**

Water vapor also differs greatly from other GHGs because the atmosphere can hold only so much of it. When you watch a weather forecast, you hear the term *relative humidity*, which refers to the amount of water vapor currently in the atmosphere compared to how much the atmosphere can hold. On a really hot and sticky day, the relative humidity may be 90 percent — the atmosphere has just about taken in all the water vapor it can. When the relative humidity reaches 100 percent, clouds form, and then precipitation falls, releasing the water from the air.

Ozone depleters

Chlorofluorocarbons (CFCs) are also considered GHG, responsible for about 12 percent of the greenhouse effect the planet is experiencing today. You don't find much of these CFC gases around anymore because the Montreal Protocol of 1989 required countries to discontinue their use. CFCs break down the *ozone* — the layer throughout the stratosphere that intercepts the sun's most deadly rays. (Without the ozone layer, the sun's ultraviolet rays would kill all living things.)

CFCs were mostly used in aerosol spray cans and the cooling liquids in fridges and air conditioning. The Montreal Protocol was a fantastic success. The use of chemicals that destroy the ozone layer is illegal — globally. And now the ozone layer has started repairing itself! A side effect of banning ozone-eating chemicals is that a lot of them were *also* GHGs. So protecting the ozone layer also helped the climate!

TECHNICAL
STUFF

Because CFCs are already regulated under the Montreal Protocol, they're not regulated under climate agreements. The recent Kigali Amendment to the Montreal Protocol will remove a significant amount of GHG that are also ozone depleters (Read more about what gases are covered under the Kyoto Protocol in Chapter 11.)

RUNNING UP EMISSIONS WITH YOUR SNEAKERS

Some of the sources of these gases are really wild. Here's one: Nike came out with the popular Nike Air shoe, a running shoe with a cool little air-filled bubble in the heel, in the late 1980s. That bubble was filled with — you guessed it — a GHG (Sulfur hexafluoride, to be exact)!

The amount of GHG in those shoes all together added an equivalent of about 7 million metric tons of carbon dioxide — or the emissions from 1 million cars — into the air when they hit the garbage dump after being worn out. In the summer of 2006, after 14 years of research and pressure from environmental groups, Nike stopped using the GHG in shoes and replaced it with nitrogen. We're glad that bubble burst.

Unfortunately, getting sulfur hexafluoride out of Nike runners didn't eliminate all sources. Sulfur hexafluoride levels in the atmosphere continue to rise due to its use in the electricity sector. Concentrations of sulfur hexafluoride have been creeping up by about 0.36 parts per trillion (ppt) per year. In 2021, it made up 10.66 ppt of our atmosphere.

» Linking carbon dioxide to
   temperature trends

» Understanding what happens when
   the temperature becomes too hot to
   handle

» Limiting greenhouse gas emissions

Chapter **3**

Recognizing the Big Deal about Carbon

E lizabeth and Zöe Caron wrote *Global Warming For Dummies* in 2009. Back then, the consensus of the world's scientists was that

» Climate change was caused by rising carbon and other greenhouse gases (GHGs) emitted to the atmosphere ss a result of human activity.

» Climate change could become irreversible and catastrophic if emissions kept increasing.

» The use of fossil fuels had to be reduced drastically.

The solid worldwide scientific consensus has only been reinforced since then. And now the consensus isn't just among scientists — it has become common under-standing. Humans are having real-time experiences of floods, fires, droughts, heatwaves, and other disasters. The worst-case projections for the speed and magnitude have turned out to be not worse enough.

Still, even though the science is settled, scientists still say global warming is a theory. A scientific *theory* is based on a set of principles that describe a particular phenomenon — Newton's so-called "law of gravity" is properly called his "theory of gravity." Theories aren't technically facts, but sometimes theories become so strong that people accept them as facts.

But how do you know this theory of GHG emissions causing climate change is correct? Can you really trust all those bigwig scientists? And if it's correct, what does this theory suggest is going to happen next? We answer those questions in this chapter.

Considering Other Causes of Global Warming

Sure, some uncertainty exists around exactly where the global climate system will hit tipping points in the atmosphere, how global changes will affect local weather, or how much of it is due to human activity. What if GHGs aren't the only culprits behind climate change?

The climate is an incredibly complex system affected by the sun, cloud cover, and complex long-term trend. In this section, you read about possible other drivers for climate change and see why no other has nearly the effect of GHG emissions from burning fossil fuels.

The Intergovernmental Panel on Climate Change (IPCC) says that it's 99 percent certain that increasing carbon dioxide levels due to human activity are the major cause of climate change. The IPCC says that there's less than a 1 percent chance that climate change is being caused by natural factors.

So, although some uncertainty exists, the debate on the human contribution is largely settled. With that said, some natural climate change drivers need discussion. That's largely because, over the years, these natural climate change drivers have been advanced to counter the growing consensus that human activity, primarily by burning fossil fuels and cutting forests, is the primary cause of dangerous changes in our global climate system. Putting the other natural culprits in the following sections aside is key to focusing on what humanity must do to preserve a habitable planet.

Solar cycles — Irradiance and Milankovitch

The sun has different cycles, and the Earth's climate changes over time in response to these cycles. The sun goes through *irradiance cycles,* when the amount of solar radiation reaching the Earth varies. Scientists were only able to measure these cycles with precision after 1978 and the advent of satellites. Two of these cycles seem to exist, one running for 11 years and the other running for 22 years.

The solar cycles do affect climate in the short term, but the U.S. National Oceanic and Atmospheric Administration (NOAA) reports that the impact from the light intensity of the sun versus the impact from GHG emissions is a ratio of approximately 9 to 40. So, GHGs have more than four times the effect of solar cycles.

Other cycles that concern the sun are the Milankovitch cycles. Although they sound like Mr. Milankovitch's bike collection, *Milankovitch cycles* are actually natural cycles of the Earth — one of these cycles, for example, is the way in which the planet tilts toward or away from the sun. These cycles may explain the *glacial cycles* — the ins and outs of ice ages. (We talk about the Milankovitch cycles in more detail in the section "Making the Case for Carbon," later in this chapter.)

Although they're important, the Milankovitch cycles have minimal effect on climate in comparison to the effects from GHG emissions when you look at them in terms of relatively short timescales — from decades to centuries. Overall, the IPCC says that the sun likely has little to do with global average temperature rises since 1950. In fact, computer models suggest that the Earth would be cooling if not for increases in GHGs.

Cloud cover

Scientists have known for a long time that climate affects rainfall. NASA (the National Aeronautics and Space Administration in the United States) has shown that the relationship may work in reverse, however — that the changing rain patterns might, in turn, indirectly affect global warming. Rainfall patterns correspond to cloud cover. Depending on their thickness and shape, clouds can reflect light during the day and hold in surface heat overnight (see Figure 3-1). The amount of water vapor in the air has continued to increase (which we talk about in Chapter 2), which means more clouds, which means more rainfall. Perhaps this increase in cloud cover can explain why nighttime temperatures are rising more than daytime temperatures in global warming trends. It gives a whole new meaning to having a hot night!

Ultimately, however, increased cloud cover seems to be a result, not a cause, of climate change. But, like the increased water vapor, it may further contribute to global warming.

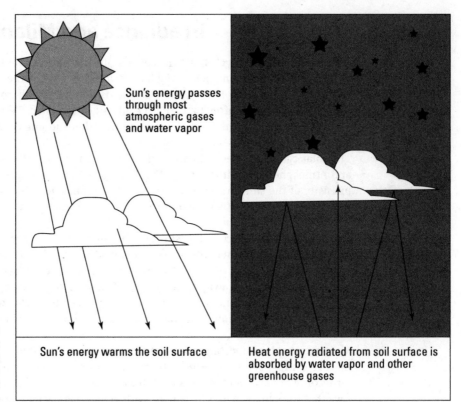

FIGURE 3-1:
Clouds reflect light during the day and hold in surface heat overnight.

Sun's energy passes through most atmospheric gases and water vapor

Sun's energy warms the soil surface

Heat energy radiated from soil surface is absorbed by water vapor and other greenhouse gases

Long-term climate trends

Over the course of many millions of years, the Earth's temperature has varied widely. Scientists know that the most recent period of ice ages started about 2 million years ago, and since 800,000 years ago, the planet started into a cycle of an ice age every 100,000 years or so. Currently, the Earth is in an interglacial period — meaning the weather is warm and stable enough that humans can develop and expand societies. Human civilization started at the beginning of this interglacial period about 10,000 years ago. Given that past warm interglacial periods lasted about 10,000 years, on average, scientists would expect the world to be getting cooler, not warmer. In fact, it appears that this cooling was happening between the middle ages and the 19th century (the little ice age), but then came the Industrial Revolution.

EL NIÑO: GLOBAL WARMING CAUSE, EFFECT, OR BOTH?

El Niño is a natural weather cycle that has the power to change global temperatures. It's been around for hundreds — possibly millions — of years. El Niño involves the tropical Pacific Ocean warming by 0.9 degrees F (0.5 degrees C) or more for about three months at a time. This warmed water eventually loses that heat to the atmosphere, causing the average air temperature (at the surface, or where humanity lives in the lower part of the atmosphere) to go up a few months later, which then alters the overall climate temporarily. The temperature of the ocean then settles back down to normal and returns to its regular cycle of ups and downs.

Scientists don't yet know whether global warming is affecting these cycles, but global warming and El Niño cycles are very interrelated. Part of the reason scientists can't distinguish between the impacts of global warming and El Niño on climate is because they're so linked and because they both influence many different aspects of regional climate that they can actually change one another. A computer model giving future scenarios of climate that includes both El Niño and global warming doesn't yet exist because of the difficulty that exists between identifying the separations between the two.

Some models say El Niño will become stronger, but others say it'll weaken. Evidence suggests that El Niño cycles have been stronger and happening more often over the past few decades, and climate models project that climate change will cause sea-surface temperatures to rise in the tropical Pacific Ocean — similar to El Niño conditions. Scientists are working constantly to advance their understanding of the relationship between climate change and El Niño.

Two unknown questions remain: How much temperature rise is the result of El Niño and how much is the result of global warming? Are El Niño temperatures any higher because of global warming?

Making the Case for Carbon

Scientists have collected evidence that confirms the buildup of carbon dioxide in the Earth's atmosphere as by far the most likely cause of the current climate crisis. And they've built computer models to forecast future effects.

Geologic and prehistoric evidence

They can measure exactly how much carbon dioxide has been in the Earth's atmosphere historically. Climatologists (scientists specializing in climate science) have drilled deep — as deep as 2 miles (3 kilometers) — into ancient ice in places such as Antarctica and Greenland. They've pulled up ice cores — long, thin samples of many layers of ice that has been packed down over thousands of years, which look like really (really) long pool noodles. (Figure 3-2 shows two scientists drilling for an ice core sample.) When the layers are clearly visible, the ice core looks like a pool noodle with horizontal stripes.

FIGURE 3-2: Drilling for an ice core sample.

National Oceanic and Atmospheric Administration

Scientists can date an ice core by counting the layers of ice — just like you can tell the age of a tree by counting its rings. The layers of ice tell them exactly when the ice was formed. Each layer of ice includes little pockets of trapped air. These frozen air bubbles are like time capsules of the ancient atmosphere. They're full of gas, including carbon dioxide, that has been trapped for hundreds of thousands of years. Each layer of ice in the ice core also contains *deuterium*, a hydrogen isotope that enables scientists to determine what the temperature was when that ice layer was formed.

TECHNICAL STUFF

An atmospheric temperature change of just 1.8 degree F (1 degree Celsius) leads to a change of 9 parts per million (ppm) in the amount of deuterium stored in the ice. By contrasting the ancient temperatures revealed through the analysis of the layer's deuterium and carbon dioxide, scientists can glimpse the relationship between historical levels of carbon dioxide and temperature. The two run side-by-side almost like the lanes of a race track.

Scientists still don't know the exact cause and effect relationship between GHGs and temperature throughout the planet's history. The cause of the last ice age, for instance, probably wasn't a drop in atmospheric carbon dioxide, but a result of the Earth tilting away from the sun in a phase in the planet's Milankovitch cycle (which we discuss in the section "Considering Causes of Global Warming Other Than GHGs," earlier in this chapter). This cooling then spurred the atmosphere's carbon dioxide to drop, and the two events in tandem brought about the ice age. Ultimately, scientists still aren't sure whether temperature affects carbon dioxide, or whether carbon dioxide affects temperature — it's a question of which came first, the chicken or the egg.

What scientists do know for certain is that a distinct pattern and relationship between carbon dioxide and temperature exists; when one is high, so is the other, and when one is low, the other plunges, too. Scientists also know that the Milankovitch cycle has little to do with climate change over the past 200 or 300 years. In that time, human-produced carbon dioxide levels have skyrocketed, and temperature is starting to follow. As a result, scientists are certain that human-produced GHGs are currently warming the Earth. This close relationship between GHG concentrations and temperature suggests these higher levels of carbon dioxide will cause temperatures to continue rising.

Figure 3-3 shows the historic connection between carbon dioxide concentrations and fluctuations in temperature, as captured in ice-core deuterium levels.

Modeling and forecasting

To look forward, scientists make climate models (a *model* in this case is a computer program) to simulate the functioning of the Earth's atmosphere and climate. These models cover the atmosphere, oceans, land, and ice of the planet. Researchers input data about the climate and how it works, and then start modifying that data to create various alternative scenarios.

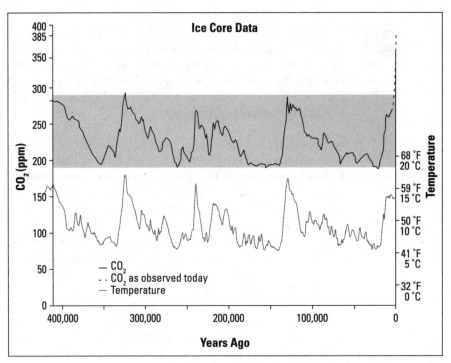

FIGURE 3-3:
GHG levels and temperature fluctuations over the past 420,000 years.

Source: National Oceanic and Atmospheric (NOAA). Vostok Ice Core data and Mauna Loa carbon dioxide observations. Graph: John Streicker.

These models have been proven; they work. Even conservative financial newspapers not known for positive views on climate change admit that the models have been "largely correct" in indicating the future path. The models keep getting more comprehensive and detailed — now people can see with real clarity the effects of continuing to emit carbon into the atmosphere.

The modelers are able to look at what would happen if, say, temperatures went up by, say, 3.6 or even 10.8 degrees F (2 or 6 degrees C) above 1850 levels. (These temperature increases refer to the global average, which we discuss in the following section.) (See the "How climate models work" sidebar, in this chapter, for more information.)

In particular, the models show how sensitive the climate is to what may seem like very small changes in temperature.

HOW CLIMATE COMPUTER MODELS WORK

The climate is affected by both the atmosphere (the part that everyone talks about the most) and the oceans. Changes in the air happen quickly, and changes in the oceans happen very slowly. So, scientists have been able to study air changes relatively easily, but they have quite literally had to wait and see what happens to the oceans. And because the ocean actually affects the bulk of the climate, they're also having to wait and see what happens to the entire climate. So, scientists need climate models, projected scenarios created by super computers, to help predict major climate changes.

The most complex climate models, such as those used at NASA, look at the Earth in three dimensions. The scientists divide the atmosphere and oceans into square columns for input to the computer models. Each of these columns has its own set of weather information based on the history and current status of the area. This information gives the computer a base to work from. Then, the researcher running the model changes the numbers to see what would happen if one condition changed, such as air temperature. For short-term projections (looking forward a day to a month), an advanced computer can make the calculation in 20 minutes. But making longer-term projections (such as 50 years from now) can take a month or two. A global circulation climate model can take as long as a year to produce results after researchers input all the variables.

A few degrees is a lot

Three or four degrees F seems like a small number to make a big deal about. You may even be thinking that an extra 3.6 degrees F (2 degrees C) seems like a perfect amount of global warming. Your garden would grow better, you'd be hitting the beach more often, and the golf season might be longer, right? But 3.6 degrees F (2 degrees C) is actually a lot. The IPCC reports that the global average temperature in the middle of the last ice age was only 10 degrees F (5.5 degrees C) colder than it is today.

REMEMBER

This increase of 3.6 degrees F refers to the average global temperature, but average numbers hide the extremes on either end. For example, you can dive into a pool that has an average depth of 1 foot (30 centimeters) if it's 10 feet (3 meters) at the deep end. Right now, the average global temperature is 60 degrees F (15.6 degrees C). Of course, temperatures can be much colder than that in the winter and way warmer in the summer. In that same 60-degree Fahrenheit (15.6-degree Celsius) global average, you can go skiing in the Alps or swimming in the Caribbean.

Going, going, gone . . . The tipping point

The *tipping point* is the point at which something has gone too far — or past the point of no return. Think of slowly going up the first climb on a roller coaster. After you go over the top, no one can stop the ride.

REMEMBER

Scientists believe that climate change has a tipping point, when the damage becomes too great to be reversed. After this point, not only can nothing reverse the impact on the planet, but little could stop that impact from increasing, either — it just keeps getting worse.

Scientists figured out, for example, how hot the climate would need to become to melt the entire world's ice sheets — this melt would cause sea levels to rise, which would flood coastal cities around the world. At the same time, the scientists figured out the amount of GHGs needed to reach these temperatures.

By looking at these different possibilities, scientists can tell which effects of climate change humans can deal with and which ones are beyond humanity's ability to adapt to or control.

WARNING

In *Global Warming For Dummies*, we said that that IPCC had defined an average global temperature rise of about 3.6 degrees F (2 degrees C) above 1850 levels as the official climate change warning zone. But scientists know now that that temperature increase is above the point of no return, or the tipping point. In 2018, in the IPCC's Special Report on 1.5 Degrees, it found that even holding to 1.5 degrees is no guarantee of a safe climate. If the temperature goes to 5.4 degrees F (3 degrees C), then 7.2 degrees F (4 degrees C) becomes inevitable. At 7.2 degrees F (4 degrees C), then 9 degrees F (5 degrees C) becomes inevitable. And so on. The increases soon outstrip any human ability to slow or control those increases. The increased warming becomes inevitable because of positive feedback loops (see Chapter 7 for the lowdown on feedback loops). Thawing *permafrost* (the frozen ground in the Arctic) releases methane, speeding more warming. Melting icecaps reveal more open water, which is darker than snow and absorbs rather than reflects the heat of the sun, speeding warmer ocean temperatures and more ice melt. Dryer conditions lead to more forest fires, releasing more carbon and causing more warming. This domino effect can lead to an unlivable world.

REMEMBER

No one knows exactly where that tipping point for global warming is. Scientists know only that humanity has a chance to avoid it by holding to 1.5 degrees C. The IPCC says that the average global temperature will rise by 3.6 degrees F (2 degrees C) when total carbon dioxide levels reach 425 to 450 ppm. The trouble is, it's moving upward at 2 ppm per year and is currently at about 412 ppm of carbon dioxide as we write this.

Uncertainty always exists when it comes to making predictions. The IPCC has found that "The scale of recent changes across the climate system as a whole and the present state of many aspects of the climate system are unprecedented over many centuries to many thousands of years." The principles in the original 1992 treaty, the UNFCCC, called for action even if uncertainty remained: It's better to be safe than sorry. The United Nations calls this philosophy the precautionary principle. Unfortunately, governments didn't exercise the "precautionary" actions. That's why the Earth is already in a danger zone.

Eyeballing the Consequences of Continued Carbon Dioxide Increases

If scientists are right about the connection between carbon dioxide and climate change, then what comes next? The past is all very interesting, but it's history. What the future holds concerns all of humanity — and the predictions that scientists have for the future are alarming.

As the ice cores demonstrate, carbon dioxide levels have always fluctuated (check out the section, "Making the Case for Carbon" earlier in the chapter for more information), but the atmosphere now has 35 percent more carbon dioxide than at any time in the last 800,000 years. Historically, carbon dioxide has reached highs of 280 ppm at a maximum. The atmosphere is now at 412 ppm.

This increase in carbon dioxide is an extraordinary shift. If present trends continue, the Earth's average temperature is likely to be 3.6 to 10.8 degrees F (2 to 6 degrees C) above 1850 temperatures — and that temperature increase could be disastrous for all life on Earth. The Earth's temperature has already risen approximately 1.4 degrees F (1.11 degrees C). You find more about what a rise of 1.5 degree or 2 degrees C would mean in Chapters 10 and 17.

WHY SCIENTISTS COMPARE TEMPERATURES TO THE YEAR 1850

The international scientific community uses the temperatures at the time just before the Industrial Revolution (1850) as a baseline. They do so because human contributions to climate change weren't significant before that time. By measuring the buildup of GHGs and temperatures compared to what they were before the Industrial Revolution, they're measuring the impact that is largely attributable to human activity.

What happens when the mercury rises

In 2009, the earlier version of this book *Global Warming for Dummies* talked about the need to keep the world's average temperature below 3.6 degrees F (2 degrees C). That goal is now understood to be more dangerous than was originally thought. Scientists now confirm that the humanity should strive for no more than 1.5 degrees C or anticipate the following:

>> Increased droughts in what are now semi-arid areas

>> Loss of food production from persistent drought

>> Dislocation of hundreds of millions of people due to climatic events with potential global political instability

>> Rapid Arctic ice melt

>> Fast loss of permafrost and the release of methane

>> Increased forest fires and wildfires

>> Shifting air circulation globally, such as collapsing *polar vortex* (a large long-lasting low-pressure area at both the north and south poles), increasing dangerous events from heat domes to extreme cold in southern latitudes

>> Increased damage to cities and major infrastructures because of higher-intensity storms and floods

>> Doubling the risk of multiple species extinctions

Figure 3-4 outlines the changes that different temperature increases, up to and beyond 3.6 F (2 degrees C), may bring. For more on the consequences of global warming and other climate changes, check out Part 3.

Cutting back on carbon

Modern civilization probably won't stop producing GHG emissions altogether. But to stop levels from passing much farther toward 420 ppm — and thus limit the world to a 2.7-degreesFahrenheit (1.5 degrees C) temperature rise — emissions must decline rapidly. The IPCC has laid out that the only pathway to holding to a safer temperature rise of 2.7 degrees F (1.5 degrees C) requires a 50 percent cut in carbon dioxide below 2010 levels by 2030. (See Chapter 11 for more about international climate change agreements.)

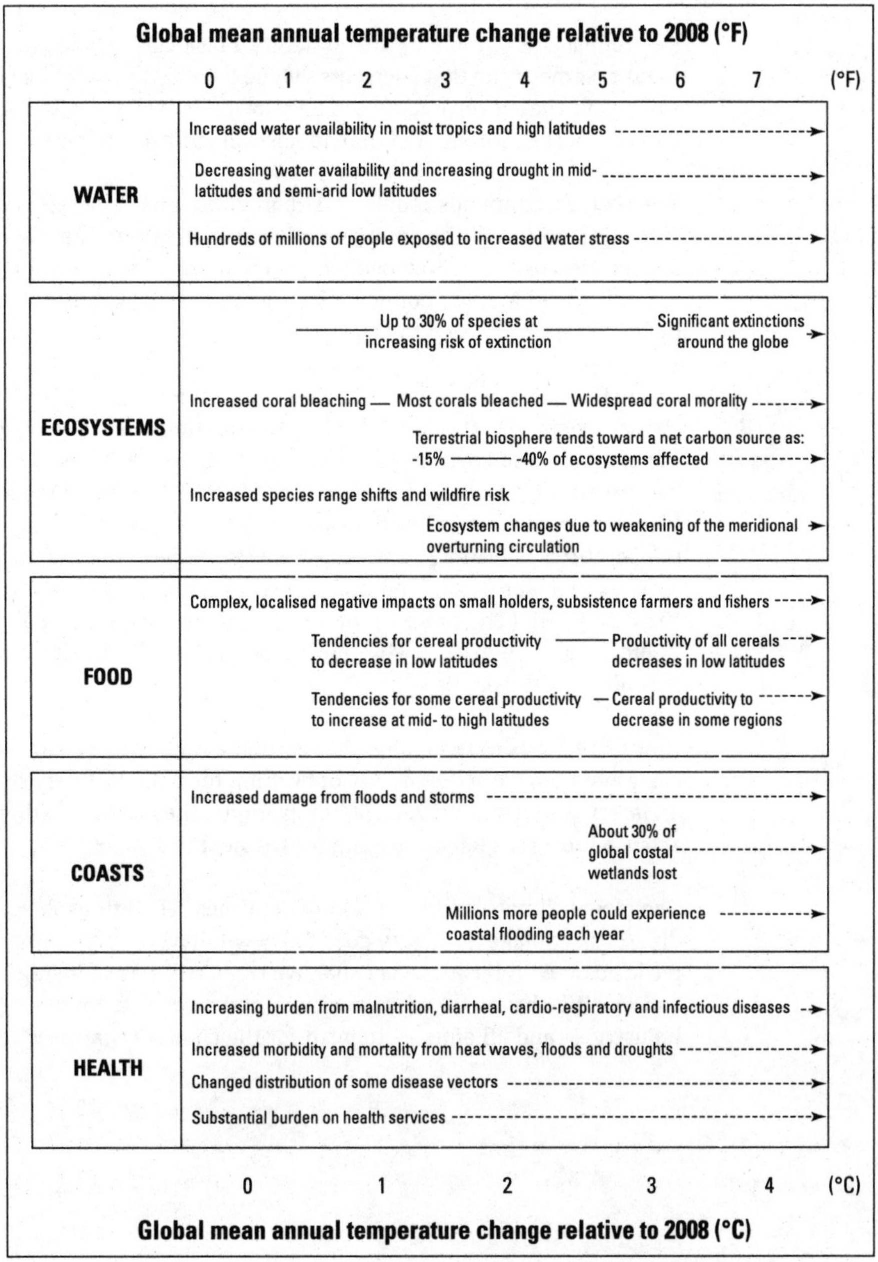

Global mean annual temperature change relative to 2008 (°F)

| | 0 | 1 | 2 | 3 | 4 | 5 | 6 | 7 | (°F) |

WATER

Increased water availability in moist tropics and high latitudes ----------------➤

Decreasing water availability and increasing drought in mid- ----------------➤
latitudes and semi-arid low latitudes

Hundreds of millions of people exposed to increased water stress -----------------➤

ECOSYSTEMS

_____ Up to 30% of species at _____ Significant extinctions
increasing risk of extinction   around the globe ➤

Increased coral bleaching — Most corals bleached — Widespread coral morality -------➤

Terrestrial biosphere tends toward a net carbon source as:
-15% ———— -40% of ecosystems affected ----------➤

Increased species range shifts and wildfire risk

Ecosystem changes due to weakening of the meridional ➤
overturning circulation

FOOD

Complex, localised negative impacts on small holders, subsistence farmers and fishers ----➤

Tendencies for cereal productivity ———— Productivity of all cereals ----➤
to decrease in low latitudes   decreases in low latitudes

Tendencies for some cereal productivity — Cereal productivity to -------➤
to increase at mid- to high latitudes   decrease in some regions

COASTS

Increased damage from floods and storms -------------------------------➤

About 30% of
global costal -----------➤
wetlands lost

Millions more people could experience -----------➤
coastal flooding each year

HEALTH

Increasing burden from malnutrition, diarrheal, cardio-respiratory and infectious diseases --➤

Increased morbidity and mortality from heat waves, floods and droughts ----------------➤

Changed distribution of some disease vectors --------------------------------➤

Substantial burden on health services ---------------------------------➤

| | 0 | 1 | 2 | 3 | 4 | (°C) |

Global mean annual temperature change relative to 2008 (°C)

FIGURE 3-4:
Effects from climate change will intensify as temperatures rise.

Modified and based on Figure 3.6. Climate Change 2007: Synthesis Report.
Fourth Assessment Report. IPCC. Cambridge University Press.

Deep emissions reductions are humanity's best choice because it's the safest route to take. Some argue that such cuts will be too disruptive to the economy. (We consider ways that governments can help lower emissions in Part 4, and in Part 5, we look at how businesses and individuals can cut back on emissions.)

The IPCC recommends reducing carbon emissions to net zero by 2050, on top of cutting them in half by 2030. Many countries and groups have committed to an aggressive 2050 goal independently, including the European Union, California, and the World Mayors Council. (See Chapter 10 for solutions being implemented by governments around the world.)

REMEMBER

The main reason a major reduction in GHG emissions is so important is because a chance exists that the planet's climate situation could get worse than predicted. For example, the Earth is experiencing changes that speed up the warming cycle because of positive feedback loops (see Chapter 7 to find out about these feedback loops), or because parts of the carbon cycle are weakening because of increasing temperatures, meaning that not as much carbon is being sucked up by carbon sinks such as forests and oceans — leaving humans to deal with more emissions than expected. (The precautionary principle, which we talk about in the section, "Going, going, gone . . . The tipping point," earlier in this chapter, looks more appealing by the day!)

Humanity needs to remember that we don't start with a clean slate every year. We carry a burden from the history of burning fossil fuels. Every time a person drives an internal combustion vehicle, for example, they release carbon dioxide that will act as a force for global warming for the next 100 years.

The global climate system has long lag times. The atmosphere doesn't turn on a dime. The damage humanity does today will have an effect over a century. If people keep on with business as usual, we run a real risk of losing a functional civilization. We have to stay positive and work to move governments and industry — and all of us — to push for the changes that need to be made.

2
Tracking Down the Causes

Of the many factors causing climate change recognize one substantial source of greenhouse gases stands out: fossil fuels. Explore where fossil fuels come from and why they're causing so much trouble.

Look more closely at how major industries, from manufacturing to logging to farming, are contributing to climate change.

Examine how everyone is unwittingly contributing to the problem.

Chapter **4**

Living in the Dark Ages of Fossil Fuels

D epending on where you live, you may be able to meet your energy needs without fossil fuels. For the last 150 years, that wasn't the case. People, especially in industrialized wealthy countries, met most of their energy needs by burning fossil fuels, such as coal, oil, and natural gas.

Burning these fossil fuels released large amounts of greenhouse gases (GHGs) (we talk about those gases in Chapter 2). In fact, a little more than two-thirds of human-produced GHGs in the atmosphere come directly from burning fossil fuels. In this chapter, we examine the types of fossil fuels, look at how people use them in their day-to-day lives, and assess fossil fuels' overall contribution to climate change.

From Fossils to Fuel — How Fossil Fuels Came to Be

A lot of people know that fossil fuels pollute and produce carbon, but they don't understand why. To understand that, you need to know where fossil fuels, such as coal, oil, and natural gas, come from. They're literally derived from fossils of past living matter.

Talk about fossils, and the first things that may come to mind are dinosaurs. But when it comes to the fossils in fossil fuels, they're actually fossils from *before* the time of the dinosaurs — starting off as decomposing plant material (not decomposing dinosaurs).

Many of these plants grew in swamps that used to cover even the northernmost parts of the globe 300 to 400 million years ago. Usually, plants and trees rot away into the soil, but swamps don't have enough air (it's what scientists call an *anaerobic environment*) for the usual decomposition process to happen. Instead, over time, these dead plants and trees sank to the bottom of the swamps where they eventually turned into peat. The peat was buried and compressed under layers of sediments such as sand and silt. As these sediments turned into rock, more pressure was piled on the peat below it. The moisture was squeezed out of the peat like water squeezed out of a sponge, turning the peat to fossil fuels. So, millions of years later, fossil fuels are typically found deep underground.

Similar fossil fuels are also found under the ocean, where sea plants and old shells were buried and pressed down under the ocean sediment. See Figure 4-1 to get an idea of what this process looked like.

Not all plant matter in those ancient swamps and in the oceans turned into fossil fuels. The process needed the right conditions, such as enough pressure and the correct bacteria. Although many of these plants were very different from anything known about today, they sucked up carbon dioxide from the atmosphere and gave off oxygen, like all plants still do.

REMEMBER

When fossil fuels are burned for energy, those fuels release their carbon, in the form of carbon dioxide and other gases that these ancient plants stored. (For more on why releasing carbon dioxide into the atmosphere is problematic, refer to Chapter 2.)

GHGs are released not only when these fuels are burned, but also when they're retrieved from the Earth. Extracting the fuels and processing them into their final forms requires fossil fuels, and thus produces carbon dioxide. The oil has to be taken out of the Earth, transported to a refinery, processed into a usable form, and

The Creation of Fossil Fuels

The sun is the ultimate source of energy.

Tree, plants, and animals absorb the sun's energy.

Remains of trees, plants, and animals.

Earth crushes the fossils over time and, with heat, converts the long-stored sun's energy into fossil fuels.

FIGURE 4-1: How fossil fuel is created.

© John Wiley & Sons, Inc.

transported to its final destination. Because traditional sources of oil have begun to dry up, industry has turned to sources such as Alberta, Canada's oil sands, which require even more energy to yield usable fuel. (See the sidebar "How much oil is left" for more about the possible end of oil and the sidebar "Athabasca tar sands: A sticky situation" for information about the tar sands.)

REMEMBER

Fossil fuels give a one-time-only burst of energy. Take them out of the ground and burn them, and that's it. That's why they're called *nonrenewable.* The supply of fossil fuels is limited, and after people use them up, civilization will have to wait millions of years before any more exist. That's why, no matter what, civilization will have to rely on a diversity of fuel sources to produce energy in the future. We talk about alternative energy sources in Chapter 13.

Examining the Different Types of Fossil Fuels

Coal, oil, and natural gas are all fossil fuels, but they're not all the same. They differ in how they're used, how much they're used, the GHGs that they release when they're burned, and even where they come from.

When land plants, such as trees, decomposed hundreds of millions of years ago, they pressed together into a solid form known as coal. Plants and animals in the oceans decomposed in a similar way — sinking to the bottom of the ocean, getting buried under sediments, forming peat, and eventually being compressed into fossil fuels such as oil.

Each type of fossil fuel has a different amount of carbon in it, so it puts a different amount of carbon dioxide into the air when it's burned. Coal releases the most carbon dioxide when burned, natural gas the least. In the following sections, we take a closer look at the different types of fossil fuels, starting with the largest contributor, coal, and working our way down to natural gas.

REMEMBER

As the costs of renewables continues to drop rapidly, it turns out that, rather than having enormous costs, the transition away from fossil fuels will generate net economic benefits of trillions of dollars. And rather than running out of fossil fuels, most of what's left will no longer be profitable and will be left in the ground. The IEA has made it clear that to hold to safe levels of warming, most of the known reserves of fossil fuels have to be left the ground. Chapter 18 deals with this astonishingly quick disruption of the world's energy supplies.

Coal

You may think that coal was king forever ago, but about a quarter of the world's energy still comes from this fossil fuel. Coal was the first fossil fuel that humans burned for energy — in fact, the use of coal predates written history. Here we delve deeper into what's in coal and what's been done to decrease pollution from coal.

Dissecting coal — What is it?

Coal is a very dirty fuel in terms of releasing GHGs as well as a witch's brew of nasties. Toxic substances like arsenic and mercury along with cancer-causing chemicals are all released when burning coal.

Because it's essentially carbon, it releases carbon dioxide when burned, along with many other dangerous pollutants. In December 1952, to cite one dreadful example, a massive lull in air circulation trapped the coal smoke from tens of thousands of London homes over the English city, creating a blanket of pollution. In four days, the deadly *smog* (the name comes from combining smoke and fog) killed upward of 4,000 people directly, with 8,000 more succumbing to respiratory illnesses later.

TECHNICAL
STUFF

Some of the noxious stuff inside coal includes sulfur dioxide, mercury, and a huge array of polyaromatic *hydrocarbons* (cancer-causing and hormone-disrupting toxic chemicals, also in oil and gas). And it doesn't stop there. Coal also releases arsenic and cyanide; *carcinogens* (things that promote cancer), such as benzene, naphthalene, and toluene; and a witch's brew of other nasties.

Lowering pollution from coal plants

Clean coal doesn't exist. Options for cleaner uses of coal, however, do. Over a period of decades, many technological advances have been made to reduce pollution from burning coal. The following actions were taken to diminish pollution from coal plants:

>> The first step taken to reduce pollution from coal plants was aimed at reducing nitrous oxide emissions. To do this, the coal was burned at incredibly high temperatures (around 1,500 degrees F) — this is considered a low temperature compared to the 2,500 degrees F at which coal is usually burned. At such a low temperature, nitrogen doesn't combine with oxygen, thus no nitrogen oxide is created. This process happens during the burning process and reduces many pollutants but does nothing to reduce carbon dioxide.

>> The next step in pollution reduction was when coal-fired power plants in many industrialized countries added scrubbers in the 1970s to capture the sulfur and prevent it from falling to Earth as acid rain. Scrubbers are technically called *flue gas desulfurization units* — devices installed right in the flue. The device sprays a specially made liquid mix of water and powdered limestone right into the emissions coming from the burning coal. The spray immediately soaks up and becomes one with the sulfur, trapping it in this new solid material.

>> Another way of "cleaning" coal is called *fluidized bed combustion,* where the coal actually becomes liquid in the bed of the furnace. Scrubbers and fluidized bed combustion reduce emissions of nitrogen and sulfur dioxide, but not of carbon dioxide. Industries were even able to extract the sulfur and sell it, increasing their profits. Removing sulfur dioxide was a step in the right direction for solving the problem of acid rain, but these scrubbers, again, do nothing to reduce carbon dioxide, mercury, or the whole array of other pollutants.

Research and development teams are devoting a lot of time and energy to producing a type of coal that doesn't add to GHGs. One idea suggests turning coal into a gas and stripping the carbon dioxide out of that gas, then storing the carbon dioxide in the ground. The technology to actually strip the carbon dioxide from the gas doesn't yet exist, but the carbon-storage technology does (and parts of Europe already use it). Until the day that carbon dioxide can be stripped out of coal, conservation practices and replacing coal-fired power plants with cleaner, renewable fuels are the most effective and sustainable ways to reduce GHGs. (Flip to Chapter 13 for more on clean fuels and carbon storage.)

Coming out of COP26 in Glasgow, for the first time climate negotiators from every country on earth agreed that everyone has to phase down coal use together. Many countries, including the UK and Canada, are part of the Powering Past Coal Alliance. Increasingly, in the context of a rapidly depleting global carbon budget, governments agree coal-burning for electricity must end. The search for finding cleaner ways to burn it is replaced with the race to replace it.

Oil

Today, oil provides about 33 percent of the world's energy. People use it each time they fill up their cars, get on a plane, or turn on an oil furnace. Oil is also the key raw material used for manufacturing a wide variety of very common products, including plastics (from food containers to toys), artificial fibers, and a host of other goods such as hair gels, shampoo, deodorant, and dishwashing liquid.

HOW MUCH OIL IS LEFT?

Cheap oil has been the lifeblood of the post–World War II economic boom.

Climate change activists have been urging people to reduce their fossil fuel consumption because of the impact on global warming, but another compelling reason to cut back on oil use exists: It's running out. There are roughly 1.7 trillion barrels of proved oil reserves left, according to British Petroleum reports. *Proved oil reserves* are estimated volumes of oil with an 80 to 90 percent certainty, according to the International Energy Agency (IEA). The argument about when, exactly, the world will run out of fossil fuels (particularly oil) has been going on for decades. When the Club of Rome released its famous Limits to Growth report in 1970, it said with certainty that the planet was running out of oil. The 1990s once again saw the rise of the argument that the planet would soon be out of oil. This time, the alarm was raised by geologist M. King Hubbert's concept of peak oil. Hubbert's peak referred to the point at which people would begin depleting known reserves, or when oil consumption is higher than oil production. Today's *peakers,* as they're known, are finding a lot of evidence that this point has passed. They argue not that Earth is running out of oil but that Earth has already run out of *cheap* oil.

The IEA estimates that the world economy needs to find an additional 3.2 million barrels of oil a day. Every single day, the world's petroleum geologists, and oil and gas companies, must find new sources of oil — new oil fields and new bitumen deposits equal to 3.2 million barrels of oil — just to keep the current supply steady.

The July 18, 2007, report that the U.S. National Petroleum Council (NPC) sent to the U.S. Secretary of Energy states that 80 percent of today's oil production must be replaced with new sources of oil or other energy sources within the next 25 years. Peakers argue that when the crunch hits, it will really hurt, causing recessions. People won't be able to afford the gas to fuel their cars, and suburbs will suffer. But people can see it coming and can start investing in energy efficiency and smarter ways of using the oil that's available. And using more renewable fuels could help cushion the blow of more expensive, dwindling fossil fuels. Climate change activists know that the atmosphere is running out of space for the wastes from burning fossil fuels, no matter how much longer supplies last.

Virtually everyone agrees: The age of cheap oil is over.

TECHNICAL STUFF

All petroleum products start out as crude oil or solid bitumen. A barrel of crude is more than just a barrel of crude though. You can have sweet crude and regular (or sour) crude. (Sweet crude has lower sulfur content.) Then, the light and heavy crude classifications depend on, quite literally, how light and heavy the crude is. Whatever the type, crude oil is the straight-up oil, before anyone does anything to it. Refineries process the crude oil to make gasoline for cars, diesel fuel for trains and trucks, heating oil for homes, and jet fuel used in airplanes.

When you think of oil, you probably imagine it shooting up out of the ground like a fountain. But those *Beverly Hillbillies* days of "black gold, Texas tea" are long over. Humanity has already used up most of those easy-to-tap reserves.

WARNING

Oil is starting to play hard-to-get. Although many disagree about whether humanity has hit *peak oil* — the point of maximum production of oil, after which the supply begins to be depleted (see the sidebar "How much oil is left?" in this chapter, for details). Companies are still making big oil discoveries, although a small fraction of the 200 billion barrels a year that they found in the early 1960s. Whenever the price of oil collapses, which it has done periodically since the 1980s, the level of investment in exploration drops off — so less is discovered. What is discovered these days can be enormous finds, but increasingly in hard-to-get places.

Sometimes companies need large amounts of water to push the oil out. Look at the Athabasca oil sands in Alberta, Canada, for example. The oil industry used to consider separating the oil from this thick, gooey mixture of clay, sand, water, and oil too expensive. But when the price of oil is high, the industry pursues a very expensive process to produce a very low value product. The process of physically pressing the oil out of the sand is very water-intensive and the energy returned on energy invested (EROEI) is a very low number. (Check out the sidebar "Athabasca tar sands: A sticky situation," in this chapter, for an in-depth look at this process.)

Offshore reserves have long been a source of oil — they're still under the ground, but also under the water. You can find these reserves off the east coast of Canada, in the Gulf of Mexico, and, in declining amounts, off the coast of Norway. Now, however, the search for new offshore oil fields is heading for more remote and fragile areas, such as the Beaufort and Chukchi Seas. These two diverse ocean ecosystems host thriving wildlife, on which the local indigenous peoples depend.

On the ground, there has been oil industry pressure for decades to drill in protected areas such as the Arctic National Wildlife Refuge or near the Okavango Delta in Namibia. Companies are also proposing projects in the Amazon rainforest in Ecuador, another fragile ecosystem that's also a vital part of the planet's carbon cycle (refer to Chapter 2 for more information).

No one would have considered these sources of oil a decade or so ago. But dwindling oil supplies and rising prices have changed all that. High prices, up to higher than $120 a barrel, collapsed in the last decade back down to as low as when oil hit $30 a barrel in the 1970s. In 2020, the price of oil dropped so low due to multiple factors including COVID that the price of oil actually went negative. As the world economies are recovering from COVID, the price of oil is rising again.

Natural gas

Natural gas is mostly methane, which makes it a little different than the other fossil fuels (check out Chapter 2 for more about methane). The cleanest of all fossil fuels, natural gas gives off only carbon dioxide and water when it burns. Rotting trees and plant matter release methane if the conditions are wet and airless. Natural gas can usually be found around coal beds or oil fields. Recently, gas has taken over coal in many U.S. utilities because of the dirty process of *fracking*, (which is a term that refers to hydraulic fracturing). This kind of gas can hardly be called anymore. (We discuss fracking more in Chapter 2.)

ATHABASCA TAR SANDS: A STICKY SITUATION

No matter what the name suggests, the tar sands (also called *oil sands*) aren't a tarry version of the Sahara Desert. They're boreal *forests* (coniferous forests, found between 50 and 60 degrees North latitude, across northern Canada, Russia, Alaska, and Asia, as well as Scandinavian Europe) and muskeg (a type of wetland found in boreal and arctic areas) that cover a sandy soil that's 10-percent bitumen, a viscous material that resembles tarry molasses. To get the oil, you have to squeeze this bitumen out of the sands.

The Athabasca tar sands hold 165 billion barrels of proven retrievable oil. These reserves make it the second-largest oil patch in the world, after Saudi Arabia. Reaching the bitumen involves stripping away the muskeg and boreal forest — a single mine may need more than 6,500 hectares (16,000 acres) of forest cleared, which also has bad consequence by removing more potential carbon sinks from the landscape.

(continued)

(continued)

After removing the muskeg and forest, oil companies dig the bitumen out of open-pit mines that are 245 feet (75 meters) deep. To extract the bitumen that lies even deeper, they have to pump huge amounts of water and steam into the ground to loosen it up and bring it to the surface. This process called in-situ mining uses between 2.5 and 4 barrels of water for every barrel of oil extracted, depending on how deep the bitumen lies. In-situ mining produces even more GHG than the open pit mines. This process creates a lot of wastewater, full of toxic waste, that they store behind enormous dikes. The tailings ponds of contaminated water cover 85 square miles (222 square kilometers).

The Alberta government permits the oil sands operation to use more than 1,177 cubic feet (359 million cubic meters) of water annually — twice as much as the city of Calgary, which has a population of more than a million people, uses in the same period. A 2006 report by the Canadian National Energy Board (now called the Canadian Energy Regulator) questioned whether the project's massive water use was sustainable. Many towns and communities, such as Fort McMurray, also rely on the river from which this water is drawn for their drinking water — the water that the mining uses may one day seriously stress the water source of Fort McMurray residents.

Heating up the bitumen and extracting it from the oil sands takes a lot of energy — energy that's supplied by . . . burning fossil fuels. These operations use the equivalent of a third to a half a barrel of oil for every barrel of oil produced. (Anyone see a losing cycle here?)

So, the mining industry consumes a huge amount of energy in order to produce oil, which primarily the United States buys for cars that don't have proper energy efficiency standards (California excluded!).

Canada's decision to keep expanding and developing the oil sands is an example for other nations of what not to do — while making oil development a top priority, it's impossible for Canada to decrease its GHG emissions. The report compared it to the American decision to encourage coal as a form of energy independence and to Brazil's clearing of rainforests.

Natural gas is almost pure methane by the time it reaches your doorstep. A quarter of the world's energy comes from natural gas. The Intergovernmental Panel on Climate Change (IPCC) reports show that the Earth contains more natural gas than regular oil, but that natural gas is patchy and spread out in comparison to oil, making it harder to tap into.

Because of how relatively clean it is when burned, some energy analysts have promoted natural gas as a clean-energy fuel to replace coal in power plants. But this solution may not be all it's cracked up to be. Natural gas

>> **Increasingly comes from nonconventional sources requiring fracking:** This involves fracturing of bedrock formations with water and chemicals under high pressure to release otherwise inaccessible sources of gas. It produces a huge volume of GHGs by releasing methane in the process.

>> **Is difficult to transport:** Moving natural gas involves liquefying it first, which requires a lot of energy. This liquefying process also creates carbon dioxide emissions, depending on the source of energy. (For instance, coal-fueled energy would create more emissions than hydroelectric energy.)

>> **Is potentially dangerous:** Concerns exist around possible pipeline explosions as well as the environmental damage created by gas exploration. Leaks and explosions do happen: On December 14, 2005, the community in Bergenfield, New Jersey, awoke to a tremendous explosion caused by a leaking natural gas pipeline that demolished an apartment building and claimed three lives.

>> **Has larger impacts:** Including increasing the number of (generally small) earthquakes.

Fueling Civilization's Growth: Adding to the Greenhouse Effect

Fossil fuels have been powering human development for a long time. Prior to accessing fossil fuels, the energy driving North America's economy was in the immoral form of human energy through slavery. This uncomfortable reality has never been fully acknowledged as a source of the initial wealth that led to the Industrial Revolution. Civilization has been steadily consuming more fossil fuels; consequently, more and more carbon dioxide has been pumped into the atmosphere. This section examines how an expanding world population along with growing economies had added to climate change.

A growing world population's impact

The world's growing population has been a key factor in the increasing levels of GHGs in the atmosphere. Earth's population was 1.2 billion in 1850, when the Industrial Revolution was taking place in its infancy. In the past 50 years alone, the population has doubled from 3 billion to more than 6 billion; today the population is nearly 8 billion. Even if the per capita use of fossil fuels had remained relatively stable, the amount of GHGs would have increased. And, of course, use keeps on growing.

Population growth is slowing and should level off. (The bad news is that this isn't expected to happen until around 2100 as the Earth's population reaches 10.9 billion people, according to Pew Research.) Nevertheless, the estimate of population numbers leveling at 10.9 billion is a better outcome than some growth curves that put the Earth at exponential growth to more than 12 billion. It all depends on reducing fertility rates, which all depends on improving the economic, educational, and political status of women and girls.

Growing economies also play a role

Countries don't produce carbon dioxide emissions equally. Unfortunately, North Americans are over-achievers when it comes to creating carbon dioxide emissions. One North American emits the same amount as two and a half Europeans, 20 Bangladeshis, or more than 40 sub-Saharan Africans!

Population pressure is a factor, but a growing economy also plays a large role in boosting emissions of fossil fuels. The modern world economy has been hard-wired to use them. Businesses and governments used to think that economic growth depended on using more and more fossil fuels. But then, in the 1970s, when major members of the Organization of Petroleum Exporting Countries (OPEC) drastically reduced oil exports for political reasons, oil prices jumped. As a result, governments encouraged people to use less oil — so they drove less, bought fuel-efficient cars, and practiced energy conservation. Industrialized nations took the first, tentative steps in reducing the use of fossil fuels.

But, after the mid-1980s (when oil prices dived), some old addictions took over. In the United States, for instance, the size of the average home (which needs fossil fuels to heat it) has increased by 50 percent since 1970 (though the size of the average family has decreased), and more drivers are using large, fuel-guzzling vehicles, such as SUVs. (You can read about improving home energy use and about more fuel-efficient vehicles in Chapter 18.) Countries such as Iceland and Sweden, however, switched to a renewable energy base and stayed that way.

Historically, the stronger a country's economy, the more GHGs it produces. The general rule has been that a strengthening economy means a greater consumption of fossil fuels — just look at the rapid growth of the auto industry in China, which promises to surpass the United States in production and sales. Some countries have broken that link. Sweden was the first to prove that it was possible to grow GDP while reducing the reliance on fossil fuels, but others have followed suit.

But even as the economies of developing countries grow, they still emit only a small portion of what people in industrialized countries do, per capita. They have a lot of catching up to do. When we wrote the first draft of this chapter in 2007, the world's biggest GHG polluter as a whole country was the United States. But, since then, China's emissions of GHGs have already surpassed the United States'. The total pollution from developing countries is expected to exceed the pollution from the industrialized world by 2030. (We take a look at developing nations in Chapter 12.)

A low-carbon future is essential. The IPCC says countries need to move quickly to clean energy or else the course of the climate emergency will become irreversible. It recommends that governments establish effective policies that support clean energies and wean the world off oil rapidly, cutting emissions by at least half by 2030. We talk more about government solutions in Chapter 10 and explore energy alternatives in Chapter 13.

GOOD NEWS

Some countries show that economic growth and carbon dioxide emissions aren't necessarily intertwined. By 2009, Sweden had seen 44-percent economic growth while reducing its GHG emissions to 8 percent below 1990 levels. Sweden is on track to be carbon-neutral by 2045.

Chapter **5**

Getting Right to the Source: The Big Emitters

N o one likes the blame game; pointing fingers and making accusations doesn't solve anything. When it comes to global warming, no one person, industry, or country is responsible for the build-up of greenhouse gases (GHGs). However, according to the British Broadcasting Company (BBC) only 100 fossil fuel companies worldwide are responsible for 71 percent of GHG emissions.

This chapter zeros in on where the bulk of emissions comes from: the big GHG emitters, including power producers, buildings' energy systems, industry, shipping goods, agriculture practices, and deforestation. We get into further detail in Chapter 6, looking at how individual decisions play into the bigger picture.

Power to the People: Energy Use

In an automated age, just about everything in today's civilization requires power — from the furnaces in buildings to the batteries in smartphones to the engines in the trucks on the highway. Unfortunately, that power isn't always Earth-friendly; most of it comes from fossil fuels (which we talk about in Chapter 4). Energy use accounts for about two-thirds of human-caused GHG emissions in the world.

About half of all the world's energy-linked emissions come from the Group of Seven (G7) countries: Canada, France, Germany, Italy, Japan, the United Kingdom, and the United States. The world's remaining 186 countries account for the rest. Sounds unbalanced, but it looks like developing countries are moving quickly to tip the scales. Rates of energy use are growing fastest in developing countries.

In fact, since 1990, GHG emissions, primarily from relying on coal burning for electricity, have grown fastest in India and China. In 1990, the United States was the largest polluter on Earth, responsible for 25 percent of all GHG emissions. By the early 2000s, China had taken over as largest fossil-fuel burning country. Still, it would be incorrect to say most of climate change is now due to China. Emissions on an annual basis don't displace all the pollution that has stayed in the atmosphere. Pollution from a hundred years ago — and everything in between — remains in the atmosphere and oceans, exerting a warming impact on our global climate. In other words, what are called "historic emissions" aren't history.

Fortunately, India and China are part of the Paris Agreement and have committed to cutting back. Unfortunately, emissions are still rising.

The following tells you which of humanity's industrial activities are the biggest culprit when it comes to climate change. Spoiler alert: It's the internal combustion engine and burning coal.

Producing electricity

Generating electricity produces large amounts of GHGs. Large-scale power plants are incredibly inefficient, and they essentially waste as much as two-thirds of the fuel that they use, either as heat sent up smoke stacks or through electricity lost along transmission lines.

Power plants take one kind of energy and turn it into another, electricity. Frequently, that initial energy source is a fossil fuel. Coal or oil plants, for example, burn the coal or oil to produce enough heat to boil water to generate steam. The force of the steam turns turbines, which creates mechanical energy. This process generates electricity that's delivered to buildings. Check out Figure 5-1 to see how a coal-powered electricity plant works.

The International Energy Agency (IEA) reports that 75 percent of Australia's electricity comes from coal. The United States uses coal for 19 percent of its electricity. Canada, on the other hand, is taking advantage of its surroundings and generates 67 percent of its electricity from renewable sources — mainly hydro, with increasing use of solar and wind — meaning far fewer emissions are generated to create electricity, according to Natural Resources Canada.

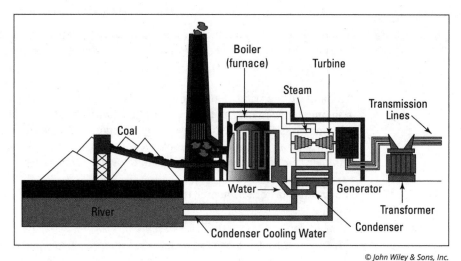

FIGURE 5-1:
Burning carbon, in the form of coal, to create electricity.

WARNING

Incredibly many countries around the world are building more coal plants primarily in Asia, despite the rise in awareness about climate change. The IEA says that coal demand dropped 4 percent through 2020 due to downturns in economic activity in the pandemic. Coal demand is rebounding post-COVID, but pressure to shut down coal is increasing. Historically, coal has been the cheapest source of electricity, but recently solar beat out coal as the cheapest way to produce electricity (Chapter 13 talks about the rapidly falling cost of renewables). As some countries continue to build coal plants, each new plant represents a 30-year commitment to infrastructure supporting burning fossil fuels and to continuing GHG emissions. (Refer to Chapter 4 for more about coal and the myth of clean coal.)

Dirty old coal plants are just that — very dirty and very old. They rely on a technology that should have left behind in the 19th century. Modern economies have more efficient forms of energy that don't need fossil fuels at all and go through far fewer steps — such as hydropower, which uses the run of the river to turn the turbines. In Chapter 13, we look at all the other ways that people can produce electricity without boiling water. (You can still make a cup of tea by using electricity flowing from a nonpolluting wind turbine or solar photovoltaic unit, though.)

Using up energy in buildings

About 15 percent of civilization's emissions come from the energy expended to heat and produce electricity for buildings, according to the International Energy Agency (IEA). Two-thirds of that energy is used in homes (see Chapter 6); commercial and institutional buildings, such as schools, colleges and universities, hospitals, shopping malls, and office buildings, use the other third.

REMEMBER

The power plants in a region or oil-run heating systems right in a building make heating and electricity possible. These plants and systems burn fossil fuels and create the GHG emissions.

Think of all the ways that people use — and waste — energy and electricity in buildings. How many times have you walked into an office building in the middle of summer and felt like you were being transported back to the Ice Age because of the extreme air conditioning? People wear suit jackets and sweaters indoors in the summer, and shirt sleeves and skirts in the winter, reversing the seasonal shifts!

Many other factors influence the GHG-producing intensity of buildings, such as whether they

>> Are well-insulated (see Figure 5-2 to see where an inadequately insulated building's heating and cooling escapes, wasting energy and creating more emissions).

>> Have proper caulking around doors and windows.

>> Make maximum use of daylight to avoid relying on electric lighting.

In Chapter 18, we talk about how you can reduce the amount of energy required to heat or cool your home.

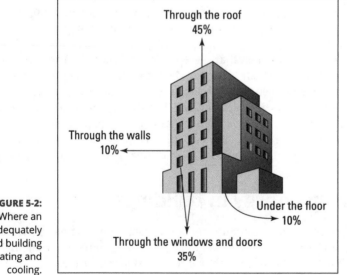

FIGURE 5-2:
Where an inadequately insulated building loses heating and cooling.

Powering industry

When people look for a single place to lay the blame for global warming, they often talk about "industry." After all, it's a pretty broad target, including just about any business that manufactures. And, admittedly, industry is responsible for a big chunk of human-produced GHGs. Just a little less than 40 percent of global carbon dioxide emissions come from energy used by industry. Industry almost exclusively releases the rest of the human-produced GHGs — including methane, sulfur hexafluoride, perfluorocarbons (PFCs), some hydrofluorocarbons (HFCs), and some nitrous dioxide. (We talk about GHGs in detail in Chapter 2.)

Industry produces GHG emissions in two main ways:

>> Burning fossil fuels to create energy and electricity.

>> Emissions that come directly from materials. The ingredients in cement production, for example, give off carbon dioxide, nitrous oxides, and sulfur dioxide.

You may be wondering how just 100 companies are responsible for 71 percent of emissions if all industry only comes to 40 percent? It's about how the statistics are sliced and diced. The first number is how many companies make and distribute the fossil fuels. The other number is about the end users for the fossil fuels. After the shift from fossil fuels to renewables is made, industry can do the same work and contribute zero carbon.

Industrialized countries can advance quickly in developing and using new technologies that have higher energy efficiency and shifting to renewable energy. Following the progress in protecting the ozone layer, after industrialized countries figured out alternatives to ozone-depleting substances, developing countries adopted those technologies at a lower cost. The same can happen as the shift is made away from fossil fuels. And now, developing countries have the chance to leapfrog straight to the newer technologies, without investing trillions, as the developed countries have, in now-outdated infrastructure. (We talk about outdated equipment and other issues surrounding developing countries in Chapter 12.)

As we discuss in the following sections, the products of industry that require the most energy are

>> Metals (primarily iron and steel)

>> Cement and other mineral production

>> Oil

>> Pulp and paper

Iron and steel

Steel is the metal that economies use most in the world. Steel-making is responsible for up to 7 percent of global human-created carbon dioxide emissions. Economists have long associated steel production with advanced industrial economies. China now produces 57 percent of all steel in the world.

Steel is manufactured in three ways:

>> **Iron ore:** This is the most common method, and unfortunately it produces the most carbon dioxide emissions. The steel is made by melting down iron ore in coke-fueled furnaces. (No, the furnaces aren't guzzling soda — *coke* is made by baking coal in airless ovens at extremely high temperatures.) Carbon and other elements are then added to the iron to produce steel.

>> **Crude steel:** You may have heard of crude steel, which, despite its name, is far more ecologically polite. It's made from the scraps from the regular steel-making process, which are melted down again in electric furnaces. This process uses about a third of the energy that the regular process does.

>> **Direct reduced iron:** The last steel-making process doesn't actually make steel at all, but makes a stand-in: direct reduced iron. Natural gas is used to melt down the iron, a production process that creates half the carbon dioxide emissions of mainstream production. (Industries in industrialized countries are trying to increase the use of this practice to modernize, with the benefit of reducing carbon dioxide emissions.)

Most steel-production emissions come from producing coke and burning coke and coal in furnaces for energy. Since the first edition of this book, major advances have been made in technology to produce low carbon steel, according to Chemical & Engineering News. Mini-mills, using advanced electric arc furnaces to recycle used steel, now make about 50 percent of U.S. steel. As electrical energy from renewable sources becomes more abundant and cheaper, it's expected that more and more steel production will be done without coal or coke.

Cement

The BBC reports that about 8 percent of GHG emitted around the world come from the making of cement. That's more than all the world's aviation!

Most of the GHGs released in making cement come from the chemical reaction in calcinating limestone to make what is called *clinker.* The rest comes from the huge kilns used to make Portland cement. These kilns use a huge amount of energy. This is one area where the good news is right around the corner.

A really exciting new technology can produce cement without producing lots of GHGs. So-called *green cement* is just as durable and strong, but actually involves changing the chemical reaction. Carbon dioxide is injected into the concrete during manufacture. This new technology is already being used at commercial scale and demonstrates where governments can regulate to ensure only green concrete is used globally.

Other metals

Producing other metals also uses a lot of energy, although people make a much smaller volume of other metals, compared to steel. Despite the smaller amount of metals produced, the industry's emissions include some of the worst GHGs, such as perfluorocarbons (PFCs), from manufacturing aluminum, and sulfur hexafluoride, from producing magnesium. Just one ton of sulfur hexafluoride, for example, has just as powerful a global warming impact as 16,000 metric tons of carbon dioxide.

Aluminum production has one of the highest rates of energy use of all industries. It requires tremendous amounts of electricity to convert raw materials to finished aluminum. Take a look at a can of soda. If the aluminum in the can came from *virgin materials* — in other words, if none of the aluminum in that can came from recycled aluminum — the energy required to produce that can is equivalent to 8 ounces of gasoline. Recycling aluminum isn't just about saving space in the landfill; it's about saving energy!

Some leading aluminum companies, such as Alcan, have been striving to reduce emissions. Alcan has reduced its emissions by 30 percent since 1990, even though its production has jumped 50 percent. Getting rid of PFCs was the key. (Flip to Chapter 14 for the scoop on industries that are implementing GHG-saving solutions.)

And not such good news, Alcan's claims have come in for revision and criticism. It was counting "reductions" when emissions were still going up! Companies often use a measure called *emission intensity* — so when emissions go down per unit of production, that's all well and good. But then when production goes up, with many more units of production — the emissions go up too. Alcan's aluminum is no longer quite so shiny.

Oil

Even though people usually talk about oil as a fuel, they sometimes forget that it's first and foremost a product. The oil industry burns fossil fuels and creates emissions when drilling for and extracting fossil fuels! In fact, 26 percent of Canada's total GHG emissions are produced in the extraction of fossil fuels. We go into all the fossil fuel details, including energy-intensive oil sand developments, in Chapter 4.

Pulp and paper

Mills that create pulp and paper products use a lot of energy. Like for so many other industries and people's homes, a power plant often creates that energy (we discuss how generating electrical power contributes to GHGs in the section "Producing electricity," earlier in this chapter).

GOOD
NEWS

Pulp and paper plants have made improvements by using wood waste to generate power and by capturing waste heat. In Canada, many pulp and paper plants generate their own electricity by using hydropower.

Pulp and paper mills also produce emissions in the paper-making processes. Mills' wastewater releases methane, and the mills produce solid waste when manufacturing the pulp and paper. And, if you want to trace the pulp and paper industry even further back, the carbon sink of trees is lost when the mills harvest those trees for paper products. (Don't feel guilty. The book you're holding is made from recycled paper and is as environmentally friendly as our publisher Wiley could make it!)

The Road to Ruin: Transportation and GHGs

Many products today originate in countries that have cheaper labor — you'll likely be hard-pressed to find products made in your own country most of the time. The computers that Elizabeth and John used to write this book were branded by North American companies, but made in China. A banana that you ate for breakfast might be from Ecuador, and your favorite T-shirt may have come from Bangladesh.

Here we explore the global connections between the stuff people buy every day and the climate crisis. Whether a hamburger or a new toaster, consumer habits have a climate impact.

Cheap goods at a high price to the climate

Today, both raw materials and finished products must travel great distances, often thousands of miles, before goods end up in the hands of consumers. And because these goods typically travel by ship, truck, or airplane (all big emitters of GHGs), most of the products that people buy and food they eat come with a hefty climate change price tag attached.

Big box stores, common throughout North America, offer low-priced merchandise that comes with a high carbon cost. These retail outlets are located where land-based tax rates are lower, and space is cheaper than in downtown areas. Such areas aren't well serviced by mass transit, meaning that shoppers must drive to the stores, causing greater GHG emissions. With shelves often stocked with goods that have been manufactured in places where environmental protections are weak — and then shipped and trucked long distances — the deals they offer aren't nearly as sweet when you factor in the climate cost of GHGs.

Keep on truckin'

Trucks have taken a larger and larger role in shipping goods in the last decade or so. And they're common — as you can tell by all the truck stops, brake-check stations, and runaway lanes that accommodate them on highways. Trucks are everywhere on U.S. highways, so it's no surprise that the International Energy Agency (IEA) has ranked the United States as the highest-producing country of carbon dioxide emissions from domestic transportation — the U.S. creates more than a third of the world's transportation emissions, a total of 1,728 million metric tons. The second-biggest producer of emissions from transportation (for now) is China, with a much lower 781 million metric tonnes.

Business used to move goods by railroad in large amounts and store the goods in warehouses. The advent of just-in-time delivery in manufacturing has increased the amount of GHGs released per product. Rather than have warehouses well-stocked with inventory that kept manufacturers and retailers ready for their work and customers, just-in-time delivery moved goods to the highways, constantly bringing goods in at the moment they're needed. Fewer warehouses, more trucks. No wonder the GHG emissions from transportation rose!

Manufacturers essentially warehouse their goods on the highways in tractor-trailers. This approach wastes a lot of energy. For example, trucks that ship frozen goods really gobble up energy. Even if the trucker stops for the night, that truck has to keep running to prevent its goods from thawing.

Companies adopt just-in-time delivery because of the money it saves them; they don't need to spend money warehousing goods. Although moving goods by truck increases congestion on highways, air pollution, traffic accidents, and (of course) GHG emissions, the companies don't have to pay for these negative side effects.

HOLDING _____ ACCOUNTABLE FOR AIR TRAVEL EMISSIONS

If you can fill in the blank, you can solve a global problem.

Air travel is tricky. It's increasing rapidly, and there are more planes in the sky every year. But international air travel was left outside international climate accords, whether Kyoto, Copenhagen, or Paris.

According to the Air Transport Action Group, a nonprofit that represents all sectors in the airline industry, government officials at international meetings still can't figure out which countries to hold accountable for the emissions from global aviation — the country from which the airplane took off or the one where it landed. The emission total is more than 915 million tons of carbon dioxide annually despite air travel being 80 percent more efficient in fuel use than it used to be. The aviation industry has been developing voluntary approaches to establish global aviation emission limits. In 2021, the U.N.'s agency for aviation unveiled a Carbon Offsetting and Reduction Scheme for International Aviation (CORSIA).

The COVID pandemic made people think differently about long supply chains through multiple national borders. This might shift economies to more local production. Even during the pandemic, people still wanted to shop till they dropped — even if the shopping was online instead of brick-and-mortar stores. Delivering all those products by air and then truck, plus the industrial production to make the goods and fuel the trucks and planes is one reason that for all of 2020 — at the first wave of shutting down travel by car and plane for much of the world — globally emissions fell by only 6.4 percent compared to 2019. The big changes in transportation are more likely to come from fuel-switching and a move to more electric vehicles — even in trucking.

Draining the Carbon Sinks: Land Use

Talking about how land use relates to giving off GHG emissions might seem a little odd, at first. After all, you don't see corn crops or clear-cut land puffing out GHG emissions like a factory smoke stack. The link between GHG emissions and land has to do with how plants and soil work together.

REMEMBER

Plants and soil are major carbon sinks that soak up carbon dioxide from the atmosphere. They take carbon dioxide out of the air and store it every day without us even noticing.

But when people cut down forests or overtill soil, carbon dioxide is released, not absorbed. In fact, between 1960 and 2010, the carbon dioxide emissions connected to land use doubled from 1 to 2 billion metric tons per year, according to the Intergovernmental Panel on Climate Change (IPCC). In the following sections, we go into exactly why land use has increased carbon dioxide emissions so drastically.

Timber! Deforestation

Destroying forests is a major human cause of increased carbon dioxide concentrations. The IPCC reports that deforestation accounts for between 25 and 30 percent of human GHG emission additions to the atmosphere, when trees are cleared for timber, farming, and land development. It stresses that the largest problems are deforestation in tropical regions and the ability for forests to regrow in temperate regions.

REMEMBER

Trees are carbon *sinks* — they suck carbon dioxide out of the air, hold onto the carbon, and release the oxygen back into the atmosphere. The fewer the trees, the less carbon dioxide gets drawn out of the air. Cutting down forests also damages the soil. Soil stores carbon, but when trees are removed, the soil becomes dryer and has more exposure to the air. When the carbon in the soil is exposed to the oxygen in the atmosphere, it produces carbon dioxide. (Check out to Chapter 2 for more information about carbon sinks.)

For many developing countries, deforestation accounts for their biggest source of GHGs. These countries clear forests primarily to create land for farming, which causes even bigger problems because agriculture is another major source of GHGs (see the following section for more on agriculture's role in GHG emissions, and check out Chapter 12 for more about developing nations).

When a rainforest goes under the knife, it has a greater impact on climate change than a northern forest. For example, farmers continue to clear large swathes of the Amazon rainforest in Brazil daily to make space for growing soybeans. As we talk about in Chapter 2, tropical rainforests are the most effective type of forest at sucking in carbon dioxide. When they get cut down, Earth loses a climate regulator. The picture gets bleaker still because these forests are often cleared by burning, which puts extra GHG emissions into the atmosphere. So, when you think globally, you have to think about deforestation as a serious climate problem. Saving the Amazon isn't just about protecting its amazing array of species. It's about saving *humanity*.

GOOD NEWS

Recent studies at www.science.org suggest that aggressive tree planting — on an almost unbelievable scale — could be a major part of the climate solution. Researchers estimate that an additional 0.9 billion hectares of suitable land could be planted with new or restored forests. Think of it as a giant carbon dioxide-sucking machine that could pull 205 gigatons of carbon out of the atmosphere. Humanity would still need to stop burning fossil fuels, but tree planting could be an important part of the response to the climate emergency.

For information about how industry can improve land management, check out Chapter 14.

Down on the farm: Agriculture and livestock

Everyone depends on agriculture to bring them daily bread — but agriculture involves more than fields of wheat and crops of corn. Food crops are a big part of the agriculture equation, but agriculture also includes growing crops for fiber, such as cotton, and keeping animals for food and other products, such as wool. The IPCC reports that 40 percent of the land that humans use on the planet is used for agricultural practices (two-thirds of that, about 25 percent, is used for grazing livestock, the rest for more intensive farming).

Almost every step in agricultural processes produces GHGs — from preparing the soil, to planting the crops, to harvesting the crops, to distributing the food. About 18 percent of world GHG emissions come from farming practices, according to the World Resources Institute.

REMEMBER

Most farm-related carbon dioxide emissions come from how the soil is treated. When it's tilled over and over and exposed to the air, the carbon stored in the soil is exposed to oxygen in the air and forms carbon dioxide.

TECHNICAL STUFF

When calculating GHGs, scientists don't count agriculture's carbon dioxide emissions. This carbon dioxide is considered *neutral* — the plants that farmers grow suck in just as much carbon dioxide as the agricultural process releases. But scientists do count how much methane and nitrous oxide agriculture produces.

The IPCC reports that agriculture accounts for half of the world's methane emissions and 60 percent of nitrous oxide emissions. Rice paddies alone account for about 8 percent of all methane emissions. Livestock, particularly cattle, represent another major source of methane — half of all human-sourced methane (32 percent), according to the UN Environment Program. People in industrialized countries still eat four times as much beef as people in developing nations, but meat consumption in poorer nations is on the rise, which means more animals are being raised, boosting emissions. Regular field crops, such as corn and wheat, produce nitrous oxide emissions, thanks to the large amounts of fertilizer that farmers use to grow them. We talk about how agriculture can produce fewer emissions in Chapter 14.

Chapter **6**

Taking It Personally: Individual Sources of Emissions

onversations about the cause of global warming typically focus on the big offenders — the worst industries, dirtiest factories, and scoff-law nations. There's nothing wrong with that. But everyone plays a role in climate change. Each of us uses energy — specifically, fossil fuels — on a daily basis.

From the moment the alarm sounds in the morning until you shut off the computer or TV at night, you're connected to an electrical grid, often fueled by coal or oil. Everyone needs energy to move from here to there — a steady supply of gas for your car or diesel fuel for the bus you ride to work. Most of humanity's food travels great distances before it arrives in homes, a journey it undertakes thanks to greenhouse gas (GHG)-producing fuels.

Typically, in talking about climate change, big business and government tend to point the finger at the individual. This is sometimes called *blaming the victim*. Asking citizens "What are *you* prepared to do?" even when it's hard and expensive to make personal changes without a major structural shift further up the food chain becomes an excuse for inaction by those with the real power to make change.

The COVID pandemic gave a real-life lesson in how much the individual is really to blame. For all of 2020, citizens everywhere around the world drove a whole lot less and flew hardly at all. You would have thought that GHGs would have dropped and in a big way. They dropped — but only by 6.4 percent, according to *Nature*. And because of emissions in previous years, the concentration of GHGs in the atmosphere continued to rise, smashing through previous records to 412.5 parts per million (ppm) — more than at any time in more than 3 million years. Still, individual choices do matter. The most powerful individual change you can make is political by letting your elected officials know you demand they take the climate emergency seriously. And the personal choices you make are important. They send a signal. They keep your own sense of personal choice and personal power intact. If you want to read more on how you can help stop global warming, check out Chapter 19.

In this chapter you can find ways to reduce your personal contribution to the climate crisis, how you get from A to B, and where the GHG emissions lurk in your home — and in your garbage can. You can make a difference.

Driving Up Emissions: Transportation and GHGs

When Henry Ford innovated the mass production line, he made ownership of the Model T a possibility for average wage earners. The Model T led to faster and bigger cars. And the car transformed the culture of North America. The technology of the car is now referred to as a *disruptive* technology. It changed the economy and society a lot — and fast! Critics often say that the big changes necessary to slow down global warming will just take too long to have real effect; it took only ten years to convert New York City from horses to cars. And that was more than a 100 years ago, when changes took much longer.

Urban landscapes shifted from focusing on impressive architecture, street cars, and pedestrian access to giving planning preference to cars. Pedestrians and cyclists weren't as important as parking lots and highways. Drive-throughs replaced the corner soda shop. Los Angeles, arguably the apex for car culture, even pioneered drive-through funeral parlors. North America's love affair with the car changed everything — including the atmosphere. Increasingly, the car was also becoming a "necessity" for large numbers of people in the developing world.

About 24 percent of all GHG emissions come from moving people and goods, according to the World Resources Institute. In the United States, the proportion is higher; closer to one third of emissions come from transportation — 29 percent.

In Canada, it's 25 percent. Almost all transportation — about 95 percent, says the Intergovernmental Panel on Climate Change (IPCC) — runs on oil-based fuels such as diesel and gas, which explains why transportation accounts for such a large portion of overall emissions (electrification of transport is rising fast around the world, but it's still only about 2.2 percent of all vehicles). Cars and diesel trucks are the top two offenders, but ships, airplanes, trains, and buses play a part, too. We discuss the culprits of driving and flying in the following sections. Figure 6-1 shows the breakdown of how each mode of transportation contributes to GHG emissions. (Refer to Chapter 5 for more information about the problems related to shipping goods.)

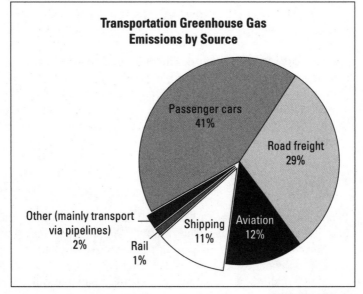

FIGURE 6-1: Passenger cars lead the way in terms of GHG emissions.

Driving

Most cars run on a simple *combustion system*. Each time you press on the accelerator, gas flows into the engine cylinder, where the spark plug ignites it. The piston goes down, and the crank shaft goes around and turns the wheels. That mini-explosion also creates exhaust fumes that get pushed out of the tailpipe. Those fumes include a dose of GHG, including carbon dioxide, nitrogen oxides, and hydrocarbons. The exhaust also contains carcinogens (such as benzene), *volatile organic compounds* (solvents that immediately evaporate into the air), carbon monoxide, and other nasties.

Whether you need to run errands or drive the kids to soccer practice, cars, minivans, and SUVs are useful and often perceived as necessary. Most households in industrialized countries own at least one car because most cities and housing developments are built around road infrastructure — making it difficult to survive without one.

However, that internal combustion engine now has serious competition — hybrid cars that generate energy to a battery to reduce fuel use and electronic vehicles (EVs), vehicles that run 100 percent on electricity. These technologies promise to make the internal combustion obsolete, just as that technology disrupted the horse and buggy. Hertz, a major U.S. car rental company, has placed an order for 100,000 Tesla EVs — that's how fast things are changing. Despite the pandemic, global sales of EVs increased by 43 percent in 2020. Still, even as sales of EVs and hybrids ramp up, in 2020 only one in every 250 cars on the road is electric.

The majority of people in the developing world still don't have access to a personal vehicle — but that's quickly changing. China is soaring ahead in private car ownership, which jumped from 45 million cars in 2009 to more than 225 million in 2019! Total EV sales in China were 1.3 million, an increase of 8 percent compared to 2019. The 2019 sales of EVs in China amounted to 41 percent of all EVs sold worldwide.

Flying

Planes burn fuel similar to kerosene, which gives off more emissions than the gasoline in your family car. The 2018 IPCC special report on 1.5 degrees found that aviation has grown to 14 percent of transport sector global carbon dioxide emissions. People made about 38.9 billion flights in 2019, although that dropped to a less than 17 billion in 2020 due to the COVID pandemic. Still, it's shocking that in 2007, globally there were 4 billion individual plane trips for business or pleasure. And, even though flights dropped dramatically in 2020 due to COVID, they have started rising once again.

The people of China are flying more and exploring their own country by air and rail. They've also increased what is called *outbound tourism* — that is Chinese tourists exploring other countries. In 2003, 20 million tourists from China explored the world. By 2019, that figure was up to 155 million! China now tops the charts for outbound tourism.

Concern about the climate crisis is leading some countries to reduce flying. Impressively, the French government decided post-COVID that no domestic flights would be allowed where train travel was available.

Increasingly, climate-aware travelers work to eliminate air travel altogether. The pandemic allowed many organizations to experience meetings — big and small — using online video technologies. The cost of flights and the wasted time traveling have likely made a permanent change in how employers see the practicality of video conferencing. This could increase the trend to staying home and avoiding flights wherever you can. If you have to fly, buying *carbon offsets* is a worthwhile option. (Check out Chapter 18 for more about green travel options.)

WARNING

Not only does flying emit a lot of GHGs, it emits them in the atmosphere in a more damaging way. The warming impact of the exhaust from air travel is far worse than the same volume of GHGs emitted on the ground.

Using Energy around the House

A person's home is their castle. An outdated saying, but the idea that a home is a castle is getting truer and truer. In Canada and the United States, the floor space of the average home has continued to grow while family size is shrinking. House size has real implications for the climate crisis. The bigger the home, the more energy required to heat, cool, and light it. Fewer people are occupying — and heating and cooling — more space.

When it comes to energy use in your home, you can think about it in two ways:

>> **Direct energy:** This term refers to the energy you use, which comes from gas or fuel oil that you consume directly — such as the oil-fed heaters or propane gas stoves you may have in your home.

>> **Indirect energy:** This term refers to how some other energy — oil, hydroelectric, wind, or nuclear power — is used to produce the direct energy. For example, natural gas is used to heat the oil sands enough so that the otherwise solid bitumen flows and can be extracted.

REMEMBER

How your electricity is produced affects your individual GHG emissions.

The energy that people use in their homes accounts for about 25 percent of GHG emissions around the world. Most of the fossil fuel energy you use directly goes toward heating your home. You use most of your electricity to power your lighting and appliances.

See Figure 6-2 for a complete breakdown of the percentages of GHG emissions produced from heating, lighting, and other energy uses.

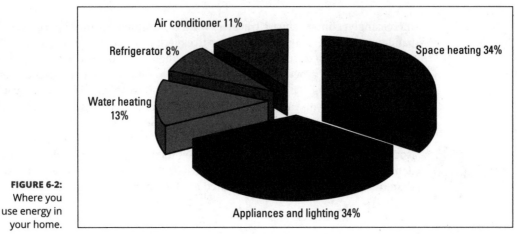

Air conditioner 11%

Refrigerator 8%

Water heating 13%

Space heating 34%

Appliances and lighting 34%

© John Wiley & Sons, Inc.

FIGURE 6-2:
Where you use energy in your home.

Your energy use may be very different than the average home. For example, you may not have an air conditioner. In Chapter 18, we look at ways to find out where you're using the most energy and electricity and what you can do to lower your energy use and reduce your dependence on fossil fuels (and save some cash).

When talking about climate change, scientists mean changes to the global climate systems. But people can also talk about "climate control" in their homes. Modern air conditioning can make the indoors feel like winter in a sweltering summer. Controlling the climate in your home can also impact the global climate system.

Controlling the climate in your home

Homes in the United States create 150 million metric tons of carbon dioxide every year from heating and cooling alone for 333 million people, according to the U.S. Department of Energy. That's a full 2020 percent of U.S. GHG emissions. Check out the following sections for how heating and cooling your home plays a role in global warming.

Heating

Heating takes either direct energy or electricity, depending on whether you have an oil or gas furnace or electrical baseboard heaters. Other types of home heating, such as wood stoves or gas fireplaces, also create emissions. The U.N Food and Agricultural Organization notes that burning wood for home heating (and in some countries, cooking, as well) accounts for about 6 percent of energy use in the world. Burning wood adds to GHG emissions both through the carbon dioxide released during burning and through deforestation.

Older furnaces emit more GHG than newer models. These old clunkers guzzle fossil fuels, but unfortunately, many homeowners cling to them, worried about the expense of buying a new unit. In reality, these homeowners can save money if they buy a new energy-efficient furnace, which would save them significantly on energy costs — and be less costly to the planet, too.

Cooling

Electricity used to be used only for keeping the lights on. Now, it's what keeps people cool all summer long. In fact, the largest share of home electricity use now goes directly to air conditioning. And in places such as south central Canada, the greater share of power demand has recently shifted from winter to summer. With more 86 degrees F (30 degrees C) days every summer — thanks to global warming — the demand for air conditioning goes up annually.

Only industrialized countries used air conditioning, for the most part, until now. Recent news reports show that sales of home air conditioners have tripled in the last ten years in China. As countries such as China and India move to catch up to industrialized countries, residents are starting to widely use luxuries such as air conditioning. Add warming temperatures into the mix, and you can see a growing air-conditioning trend and a growing demand for electricity to meet that desire.

HOW GREEN IS YOUR ELECTRICITY?

When it comes to electricity-related emissions, how much you produce depends very much on where you live. If your electricity comes from coal- or oil-fired plants, your electricity use makes a substantial addition to the atmosphere's GHGs. You can find coal plants all over the world, but they're mostly in China, India, and the United States. Some regions, however, don't create very much carbon dioxide because they generate electricity by using energy sources such as hydroelectric, wind, solar, and nuclear power. Canada and Europe have most of the large nuclear and hydroelectric plants. Germany led the way in wind power, until being overtaken by China and the United States. Spain, Denmark, and the Netherlands are also leaders in wind-generated electricity, according to the International Energy Agency.

China has led the big shift in renewable energy with by far the world's largest installed wind and solar capacity. But India and the United States are catching up. Thanks to China's big move on solar, the cost of new solar installations is cheaper than a new coal plant. That's a game changer.

Chapter 13 covers a lot of the exciting energy alternatives out there.

Traditionally, most Europeans never considered air conditioning. But because killer heatwaves have ravaged Europe in recent years, this perspective is changing. For example, the U.K. had to consider new labor laws — in the past, laws ensured a legal minimum temperature so workers could stay warm enough. Because of intense summer heat, they've also had to consider legal maximum temperatures!

Perhaps the most surprising area to need air conditioning is in Canada's far north. Buildings in the Northwest Territories and the Yukon are now being built with air conditioning. The average high temperature in the summer in those territories ranges from 70 to 80 degrees F (in the 20s C), but has been warming up recently and has reached the 90s F (about 30 degrees C).

Operating electric appliances

Think of every gadget in your home that needs to be plugged in. Actually, it may be easier to think of the things that you don't have to plug in. Every time-saving device and appliance makes you more reliant on the energy grid.

The refrigerator is the biggest electricity hog in your house — the standard fridge, purchased within the last ten years, can use 120 to 170 kilowatt hours (KWH) a month, according to Direct Energy. By contrast, a new energy-efficient fridge can use less than 350 KWH per year, less than 20 percent of what an older fridge uses.

Smaller appliances add up, too — who'd have thought that you can link toasting your waffles to climate change? Your vacuum, microwave, hair dryer, electric kettle, and coffee machine — even your toaster — all matter. Maybe they don't require all that much energy, but you use them almost every day.

Most people don't think about how much energy all today's new technology uses. The *New York Times* reports that every Google search uses 0.3 watt hours of electricity. Online streaming has replaced a lot of conventional televised entertainment, but efficiencies in the technology mean that online streaming has a minimal impact. Of course, watching on a phone uses less energy that a big screen LED.

REMEMBER

Even when you turn your appliances off, many of them are still on! The automatic instant-on features on garage door openers, televisions, and so many other gadgets pull power while they wait for you to need them. Manufacturers could make these appliances so that they demand only a fraction of the energy — and California has enacted regulations to insist they draw less power. The California Energy Commission discovered these little instant-on devices used up to as much as 10 percent of home electricity demand! If you don't live in California, your best bet is to unplug televisions, computers, and other electronics when you're not using them. (Or have them all plugged into one power bar that you can turn off with one switch.)

You Are What You Eat: Food and Carbon

Like a warm home in freezing weather, food is a necessity, not a luxury. But sadly, when people sought to make food more accessible and more convenient and to offer a greater variety, they often did so without considering the environmental toll their innovations might have.

Much of the food that people buy at the grocery store uses a lot of energy to get there — and creates a lot of GHG emissions as a result. Here are some of the key offenders:

>> **Frozen food:** Whether you're talking refrigerated or frozen, these foods burn energy when they're made, while they're being transported, and even when they're sitting in a freezer or cooler in the grocery store (or in your home). The most-energy-used-per-serving prize goes to freeze-dried coffee.

>> **Processed and packaged food:** Moving these foods through the production line takes energy, as does making the packaging (not to mention the emissions that come from all that packaging when it ends up in a landfill).

>> **Food from afar:** Elizabeth never even saw a kiwi until she was about 18 years old. Her daughter started asking for them for her school lunch in first grade. You may enjoy strawberries and mangoes in the dead of winter, when you can't pick fresh fruit right in your backyard, but moving exotic fruits and veggies around the world by plane, ship, and truck has a real cost in energy. Could people afford them if companies factored in the cost to the climate? And why should your apple be more well-traveled than you?

>> **Meat products:** Feeding livestock takes an average of 10 pounds of grain — grain that plays a large role in agricultural emissions — to produce 1 pound of meat. Also, when people eat more meat, more land is needed to raise livestock, which often means clearing forests and losing trees that breathe in our carbon dioxide. (Chapter 5 takes a look at farming.)

Wasting Away

The industrialized world isn't really a "waste not, want not" culture. It's more "Shop till you drop!" And people do shop, creating huge amounts of waste in the process.

Since World War II, household waste from the average U.S. and Canadian home has increased greatly. Over the past 55 years, people in the United States have gone from producing 2.7 pounds (1.2 kg) of waste a day to 4.9 pounds (2.1 kg) a day, according to the Environmental Protection Agency

Changes in marketing and merchandising have significantly added to packaging. Elizabeth is old enough to remember when you went to a local hardware store if you needed nails for a home building project. The person behind the counter would put the nails in a paper bag and weigh them. John has been living in rural areas where stores still have the same approach. To reduce staff and make shoplifting harder, major stores invented the dreaded bubble-pack approach. To buy a few nails, you buy a big piece of cardboard encased in hard plastic that defies any opening technique except a major attack with scissors. Multiply that packaging approach by a zillion, and you can see how society wastes so much energy in packaging and why the garbage is piling up.

About 5 percent of worldwide emissions come from waste, according to the IPCC. Landfills rot, adding methane and carbon dioxide to the air. Sewage water is another source of methane — and it adds a dose of nitrous oxide to the mix. Anything that you throw in the trash ratchets up the amount of methane in the landfill, even items such as broken furniture, old toys, and shoes. In some provinces in Canada, coffee cups make up more than a quarter of all materials in the landfill. The inventor of the single serve coffee filter now regrets the pollution load he has added to the world. Carrying your own travel mug is a good habit!

Developing countries consume far less per capita, but that doesn't mean they don't have a garbage problem. In fact, it's a different kind of problem — many countries have garbage but no funds for garbage collection or landfills, let alone the luxury of recycling plants. While developing countries are industrializing and building their economies, they're helping increase the amount of waste made in the world. Their wallets are growing, but so are their garbage piles — following in the footsteps of developed countries around the world.

The fast, cheap, and high-emission solution is to simply burn the waste. But burning waste builds up GHG emissions even further, putting us into a dizzying cycle. Because the volume of waste that civilization produces keeps rising, it has to come up with new ways to deal with it. Many great technologies enable humanity to process garbage back into what closely resembles dirt after only a few years. We go over ways to reduce the amount of garbage you produce in Chapter 18.

3

Examining the Effects of Climate Change

Get a feel for how climate change is playing a role in many of natural disasters around the world.

See how climate change is damaging other living beings and ecosystems.

Discover how climate change is affecting the daily lives of humans.

Chapter **7**

Focusing on Not-So-Natural Disasters

Major natural disasters have always happened. Storms, hurricanes, floods, and droughts are all part of the planet's natural weather and climate system.

But increasingly natural disasters aren't so natural. Human activity — burning fossil fuels and removing forest cover — has thrown the carbon balance out of whack. Greenhouse gases (GHGs) that civilization pumps into the atmosphere are driving increasingly dangerous weather (refer to Chapter 2 for more about GHGs).

Earth is experiencing more droughts, floods, more intense hurricanes, forest fires and wildfires, heavier rainfalls, rising sea levels, slowing ocean currents, and more severe winter storms and major heatwaves. For the first time, the 2021 report of the Intergovernmental Panel on Climate Change (IPCC) confirmed a risk (at low probability) that the excess carbon dioxide that people put into the air could disrupt the carbon cycle and turn the planet's life-support system into a vicious cycle.

This chapter does offer you some good reasons why civilization needs to start lowering its emissions to cool off global warming. And the good news is that, although the Earth can't go back to the more hospitable climate it once had, humanity still has time to avoid the worst. Hang on to your hat — and on to hope!

H₂ Oh No: Watery Disasters

Welcome to the blue planet. Water covers more than 70 percent of Earth's surface. And because of global warming, you might be seeing a lot more of it. Or less. It all depends on where you live.

The relationship between global warming and water is complex. Thanks to rising temperatures, some Antarctic ice is melting and raising sea levels. Elsewhere, glacial melt plays a major role in replenishing the freshwater supply to adjacent farm areas in the spring. As glaciers disappear, this water will disappear, too, increasing drought and water scarcity. Rainfall patterns are altering. Drought is becoming more common and severe. The extreme drought in the western United States through 2019–2021 was so bad that *National Geographic* came up with a new term — "megadrought."

The following sections discuss in greater detail how climate change will affect the Earth's water.

Rising sea levels

When the planet heats up, sea levels rise for two reasons:

>> Antarctic and Greenland ice caps (sitting on land) melt into the water and raise sea levels. (When the floating ice in the Arctic melts, it doesn't change sea levels but does change the water's *salinity*, or saltiness.)

>> Water expands when it warms.

Even if you take melting polar ice caps out of the picture, sea levels would still rise over the next few centuries up to about 1.3 feet (0.4 meters) for every degree Celsius (1.8 degree Fahrenheit) of temperature change.

Here we delve deeper into who rising sea levels affect and how global warming causes sea levels to rise.

Rising sea levels affect billions of people

Because the majority of the world's population lives along coasts, rising sea levels are one of the most pressing potential effects of climate change. The IPCC predicts that sea levels will rise 1 to 3 feet (0.66 to 1.11 meter) by the year 2100 from both expanding water and melting ice.

Although the predicted increase from the impact of warmer water taking more volume might not seem like a lot, it's enough to cover large parts of not only the Maldives (an island nation in the Indian Ocean, which is a scant 1 to 2 meters [3.3 to 6. 6 feet] above sea level), but also the island state of Tuvalu, 4 to 5 meters (13.1 to 16.4 feet) above sea level. Hundreds of inhabited island nations fear disappearing below the water line.

Figure 7-1 shows that sea levels were steady throughout the 1800s, but those increased carbon emissions took a rapid toll at the dawn of the 20th century. The right side of the figure shows that the pace is expected to only pick up from here.

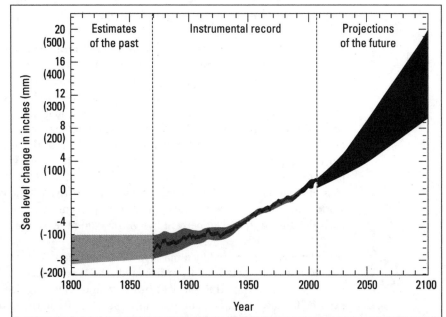

FIGURE 7-1:
Projected sea
level rise.

Source: Based on Figure 1, Section 5.1, Frequently Asked Questions. Climate Change 2007: The Physical Science Basis. Fourth Assessment Report. IPCC. Cambridge University Press.

How global warming is causing the levels to rise

Global warming is causing the sea levels to rise in two main ways:

>> When Earth's temperature rises, its water warms — and warmer water takes up more space than colder water. Essentially, the climate change is causing the ocean to expand.

>> Melting ice also plays a role in the rise of sea levels. Extra water from melting glaciers and ice sheets is flowing into the oceans.

Two areas, in particular, can potentially cause major world changes: the Western Antarctic ice sheet and the Greenland ice sheet. In the past 30 years, melting of these two massive ice sheets has increased six-fold. When Arctic ice melts, it doesn't add to sea level rise because it's ice sitting on water. Ice shelves — or *ice sheets* — are different. They sit on land. If they collapse, they displace water. "Eureka!" as Archimedes cried out. Solid objects displace water.

Surprisingly, melting sea ice doesn't affect sea levels. The water created by melting sea ice is equal in volume to the ice that was once there. Melting Arctic ice, for example, can impact the strength of ocean currents, such as the Gulf Stream, which could potentially have a serious impact on the climate, but it wouldn't cause sea levels to rise. On the other hand, if enough land ice melts, and that water makes its way into the oceans, sea levels will rise.

Consider the Western Antarctic ice sheet — an enormous body of ice that's the size of the state of Texas and contains nearly 10 percent of all the ice in the world. The on-land portion of this ice sheet appears to be weakening because *meltwater* (the water formed by the melting of snow and ice) is working its way underneath and lubricating the base of the glacier, which speeds up its slide toward the ocean. As for the ice over the water, the warming water temperatures directly under it are causing the melting. It's not likely, but it *is* possible, that this ice sheet could collapse. If that happens, the IPCC estimates that sea level rise would jump from a predicted 1.9 feet (0.59 meters) by 2100 to 13.1 to 16.4 feet (4 to 5 meters), putting coastal communities and cities around the world at risk of flooding. Many of the world's major cities are on coast lines. Forty percent of the world's population lives along a coast, and 80 percent live within 500 miles of the coast.

WARNING

Greenland is also covered by a massive ice sheet, and it's also warming far more rapidly than scientists initially anticipated. The World Meteorological Organization reports that "melting glaciers in Greenland have revealed patches of land exposed for the first time in millions of years." The ice in Greenland is more than 1.86 miles (3 kilometers) thick in some spots. Greenland's ice cover can vary from year to year, depending on the amount of snowfall and other natural weather conditions, but since the 1970s, the ice sheet has experienced a net loss of ice. NASA (the National Aeronautics and Space Administration) reports that the rate of ice loss has been increasing rapidly in Greenland. Go to `https://climate.nasa.gov/climate_resources/264/video-greenland-ice-mass-loss-2002-2020/` for a video that shows the ice loss over the past 20 years.

Melting mountain glaciers

Mountain glaciers are large masses of ice carrying rocks and dirt that usually exist at very high altitudes. Glaciers build up from snowfall over very long periods of time. In the spring and summer, glaciers start to slowly melt, and the water runs off into nearby rivers and lakes. See Figure 7-2 for a photo of a mountain glacier.

FIGURE 7-2:
Mountain glaciers
provide
freshwater.

The IPCC reported in 2006 that 75 percent of people in the world relied on fresh-water from mountain glaciers. Glaciers have been shrinking as human population has been climbing. So the percentage of available water has declined, but still half the world's population live in the watersheds of the mountainous and glaciated area called the Water Towers of Asia — also known as the "Third Pole." The region of the Tibetan Plateau through to the Himalayas feeds the seven major rivers of Asia. In India a quarter of a billion people depend on a single glacier-fed river.

Unfortunately, global warming is endangering those water sources. Glaciers are melting more quickly in the spring, releasing a lot of water at once, rather than a smaller, steady flow. This rapid melting can mean floods in the spring because glacial lakes can't hold all that water at once and drought in late summer because the water has drained away. This heavy early runoff will probably also overload the capacity of rivers and streams, eroding their banks and potentially flooding small delta areas with muddy water. Because more than 300 million people in the world live in deltas, this is a cause for concern (see Chapter 9 for more effects on people).

In addition to accelerating glaciers' melting, higher temperatures are causing most glaciers to retreat. Previously, in colder weather, snow would restore most of the ice that glaciers lost because of melting. But now, because the cold weather doesn't last as long, the glaciers don't get as much snow as they used to. Every year, more ice melts than gets replaced, so the glaciers are shrinking. Glaciers have always advanced and receded, but in the past, they did so extremely slowly (a few centimeters or an inch a year). Warmer weather is changing that. The say-ing used to be, politicians move at glacial pace. Looks like now, glaciers are mov-ing faster than politicians!

Mountain glaciers around the world are disappearing, from Patagonia to Kiliman-jaro, from the Rockies to the European Alps. For example, the 18,000-year-old Chacaltaya glacier in Bolivia, once a popular high-altitude ski destination, has completely disappeared. What tiny remnants of ice are left can't support any eco-nomic activity. The retreat of most glaciers is a clear physical indication that the world is getting warmer.

NEW OPPORTUNITIES AND CHALLENGES IN THE ARCTIC

Melting northern ice will open up new shipping routes such as the Northwest Passage, which was completely ice-free for the first time in the summer of 2007. Shipping between the Atlantic and Pacific oceans may soon become business as usual. This additional access that the new routes provide has also made oil exploration by ship much easier.

The disappearance of ice in summer in the Northwest Passage has opened up tourism opportunities. The Lindblad Explorer was the first, in 1984 bringing 103 people through the waters that were once so treacherous that many explorers, like those on the Franklin Expedition perished. In 2016, eyebrows were raised as the enormous *Crystal Serenity* brought 1,000 tourists through the legendary passage.

Arctic experts worry that the lack of infrastructure, such as ports and northern emergency services, could make such adventures unacceptably risky.

The idea that losing the Arctic ice is "good news" is disturbing given the knock on effects from melting ice. The industry excitement to obtain even more fossil fuels to speed more climate change from under the melting ice suggests a worrying denial of the seriousness of the climate change threat.

But even for those who want more oil and gas, the melting Arctic isn't all good news. The Arctic Climate Assessment Council reports that the time available each year for land-based oil exploration has been cut in half because of the warming permafrost (see the section "Examining the Negative Side Effects of Positive Feedback Loops," in this chapter, for more on the impact of melting permafrost). Open water is heating up questions of sovereignty among the countries bordering the Arctic Circle — who controls which waters, and are these newly opened areas the high seas or coastal waters? Additionally, oil exploration brings the risk of spills and other ecological disasters that would harm the fragile Arctic ecosphere.

When this Arctic ice melts, more coastline is exposed, increasing erosion. Erosion also reduces the natural barriers to storm surges. This melting ice is very bad news for indigenous communities that live in the affected areas. We discuss the effects that these major Arctic changes have on plants and animals in Chapter 8 and the effects on people in Chapter 9.

Putting a brake on the Gulf Stream

The ocean is always moving. It's not just the tides — a whole system of currents moves water in regular patterns around the world (see Figure 7-3).

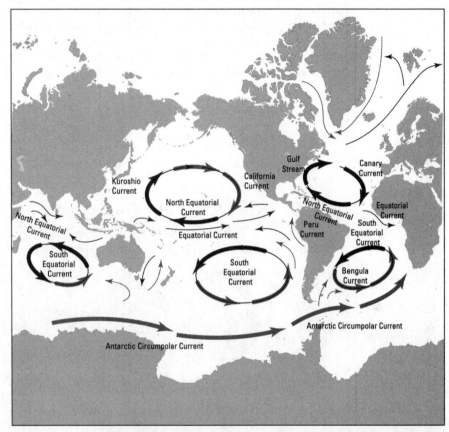

FIGURE 7-3:
World ocean currents.

Oceans have currents because water varies in density. Cold water is more dense, so it sinks, and warm water is less dense, so it floats. (When you go swimming in a lake, the top is always warmer.) Similarly, saltwater sinks, and freshwater floats. Winds move the top layer of the ocean, which then — simply by friction — moves the layers below it. This combination of temperature, salinity, and wind keeps the currents moving. Meltwater from Greenland, the Arctic, and Antarctica is affecting the ocean currents (check out the section, "Rising sea levels," earlier in this chapter for more information about this melting ice). When this freshwater is released into the oceans, it dilutes the ocean's saltiness. The less-salty water no longer sinks quickly, which can potentially slow the currents.

The Gulf Stream, located in the upper half of the Atlantic Ocean, is part of the overall ocean current system. (If you saw the movie *Finding Nemo,* it's the sea highway that all the turtles and fish cruise along.) The Gulf Stream (the technical description of the ocean conveyer belt called the Gulf Stream is Atlantic Meridional Overturning Circulation [AMOC]) brings nice warm water from the Gulf of Mexico up past northern Europe. The heat from the ocean warms the air around Europe, which helps explain why Europe tends to be warmer than Canada or Russia, even though it's at the same latitude. After the ocean releases all its warmth on Europe, it continues its route to sweep past chilly Greenland before coming over to Canada (bringing Canada cold air). Figure 7-4 shows the Gulf Stream in action.

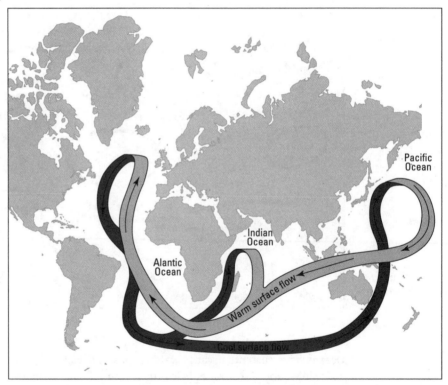

FIGURE 7-4:
Simplified flow patterns of the Gulf Stream.

© John Wiley & Sons, Inc.

Some scientists suggest that, over time, the Gulf Stream could slow or even stall because of all the extra freshwater being added to the oceans. The magazine *Nature Climate Change* recently published research findings expressing concerns the AMOC is nearing tipping points.

If the Gulf Stream slows or stops, Europe would start cooling with more severe winter storms, and rains could be disrupted to India, South America, and West

Africa, while increasing sea levels in North America. (Refer to Chapter 2 for more about the ocean's role as a carbon sink.)

Rainfall (or lack thereof)

Changes in temperature are altering evaporation and precipitation patterns, which means more rain in some places and less in others. The IPCC says these changes also mean more intense dry spells and rainstorms overall, with high-latitude areas in Europe, Russia, and Canada taking the hardest drenching.

The IPCC reports that inland mid-latitude regions — such as central Canada and inland Europe and Asia — are generally most at risk from more frequent and harsher droughts than what those areas currently experience. Although not in those regions, the land along the Mediterranean in Europe may also experience increased droughts. Droughts and high temperatures put major stress on forests and grasslands; dry, parched vegetation is a fire waiting to happen. The soil suffers, too. Dried-out soil can release into the air the carbon that it used to store. (Refer to Chapter 2 for more about how soil contributes to carbon dioxide in the atmosphere.) Drought is hard on people and animals because all living things depend on water.

Global warming is also causing deserts around the world to expand. In Africa, the Sahara is getting bigger, while in China the Gobi Desert's growth is a threat. In both places, countries are trying to arrest the increased desertification with *green walls* of tree planting. The Great Green Wall in China started with tree planting in the 1970s with billions of trees planted. Although it hasn't been entirely successful, African nations are trying to learn from mistakes in China, such as planting in all one species, and hold back the Sahara with its own Green Wall. The IPCC reports that the duo of natural warming and human-caused warming has caused the number of dry areas around the world to double since the 1970s.

Flooding

Three climate change consequences lead to flooding:

>> Rising sea levels

>> Quicker-melting snow and glaciers

>> More intense rainfall

The IPCC expects that the rising sea levels and harsher rainstorms will increase the number of floods in many places, including both *flash floods* (floods that happen very suddenly, often because of heavy rainfall and/or the ground is so dry it can't quickly absorb the rain) and *large-scale floods* (floods that stick around for a while, caused either by prolonged rainfall or water that can't drain away easily).

Climate change scenarios typically predict that average annual precipitation will remain nearly constant, but that areas will experience long periods of drought followed by an enormous volume of rain. A good example of this occurred in China in summer of 2021, when as much rain fell in three days as usually falls in a year. That nation's annual precipitation fell within days on the dry and desiccated lands. Three hundred people died. In the same summer of climate emergencies, torrential rains caused huge floods in Germany, Belgium and the Netherlands, killing 200 people.

WHAT THE IPCC MEANS BY "HIGH CONFIDENCE"

The IPCC now views the rise in increased, severe rainfall events and the flooding it causes with climate change at a level of "high confidence." Whether you're watching news reports of flooded Manhattan subways making transportation impossible, deluges creating massive damage to German villages, or torrential rains causing mudslides and killing people unlucky to be in their path, the question is no longer if the finger of blame points to global warming. It does.

Ever since the IPCC began its work in the early 1990s, it described levels of certainty in climate science in terms of confidence. That's how the IPCC instructed its authors to express levels of scientific certainty in 2018. Levels of confidence are expressed using the following qualifiers:

- **Virtually certain:** 99–100 percent probability
- **Very likely:** 90–100 percent
- **Likely:** 66–100 percent
- **About as likely as not:** 33–66 percent
- **Unlikely:** 0–33 percent
- **Very unlikely:** 0–10 percent
- **Exceptionally unlikely:** 0–1 percent

These qualifiers synthesize the IPCC author teams' judgments about the validity of findings as determined through rigorous evaluation by their scientific teams of evidence and agreement.

The most likely areas to experience more flooding are high–latitude countries, such as the United Kingdom.

Freshwater contamination

Most of the creatures that walk (or crawl or slither) on the planet require *freshwater* (water that isn't salty) to survive. Unfortunately, flooding and rising sea levels, two of the effects of climate change, pose two contamination risks to freshwater:

>> **Getting it dirty:** Runoff from flooding can get into drinking water. This runoff washes over city streets and can take anything with it — from any dirt or garbage in the streets, to overflowing sewer systems, to pesticides and fertilizers from lawns.

>> **Getting it salty:** The higher sea levels mean saltwater intrusions. Not only can saltwater get into fresh surface water, it can also work its way down into *aquifers* (water-bearing rock, which can provide well water) and coastal freshwater rivers. More people live along coastlines than any other region, and those people have the fewest sources of freshwater. Seawater contamination will only worsen that state of affairs.

Stormy Weather: More Intense Storms and Hurricanes

Global warming is heating up Earth's oceans. In fact, the IPCC reports that oceans have absorbed more than 90 percent of the heat from global warming. Hurricanes are now occurring in the top half of the northern hemisphere, such as Canada, because of these warmer ocean temperatures, particularly at the surface. Historically, colder ocean surface temperatures in the north slowed down hurricanes, turning them into powerful, but nowhere near as destructive, tropical storms. Now, however, the water's warmer temperatures don't impede storms. In fact, warming up surface water is like revving the hurricane's engine.

REMEMBER

The number of tropical storms and hurricanes hasn't increased. In fact, that number has stayed fairly uniform over the past 50 years, the IPCC reports. The intensity of tropical storms and hurricanes, however, *has* increased. Scientists have confirmed that hurricanes (also called cyclones and typhoons) have grown in strength and destructive force, while also growing in the geographical areas subject to these intense storms, moving closer to the poles. Extreme weather events from floods, to droughts, heatwaves, wildfires, and hurricanes are confirmed in the 2021 IPCC report and linked with high confidence to global warming. The first ever hurricanes to reach as far north as Canada were in 2003 in Halifax on the east coast (2003's Hurricane Juan was the first full-force tropical hurricane ever to hit Atlantic Canada) and in 2006 in Vancouver on the west.

These bigger storms and hurricanes bring rougher coastal storms, bigger storm surges, higher water levels, taller waves, more storm damage, and flooding. Scientists also see an increased tendency of these storms that stall over one area, such as Harvey in 2017, Florence in 2018, and Dorian in 2019. Some storm-protection barriers may not be strong enough to protect against the hurricanes that are coming, and some cities might need to reevaluate their protection. (Think New Orleans!)

In December 2021, a season not typical for tornadoes, northeast Arkansas, Tennessee, and western Kentucky were hit with tornadoes three-quarters of a mile wide with wind speeds that peaked between 158 and 206 miles per hour. Still, the connection between more severe and unexpected tornadoes and climate change isn't linked with high confidence — yet.

Forest Fires and Wildfires: Trees and Grasses as Fuel

As we discuss in Chapter 2, forests are critical in keeping excess carbon dioxide out of the atmosphere. Unfortunately, the number of forest fires is greatly increasing, and global warming is the cause.

The increase in hot, dry weather means drier forests, grass, and undergrowth, ideal fodder for fire. Forest fires and wildfires around the world last longer and burn with more intensity than previously recorded. The area of land burned by wildfires has surged in the past 30 years across North America. The IPCC reports that a one-degree Celsius rise (1.8 degrees Fahrenheit) in average temperatures has increased the length of the fire season in northern Asia by 30 percent. In Canada, we increasingly hear people refer to summer as "fire season."

Here we look closer at the consequences of forest fires and wildfires and ways that climate change leads to fires blazing out of control and what people can do to prevent them.

Considering the costs

These fires have serious consequences, not only for the environment, but also for infrastructure. Major wildfires in Australia, in 2020 displaced 3 billion animals, while intense fires devastated areas from California to Greece to British Columbia. The 2007 fires across the state of California destroyed 1,500 homes. As the IPCC expected, forest fires and wildfires have increased while temperatures continue to rise and some areas experience reduced rainfall.

The first major economic hit to Canada from the climate crisis was the dramatic pine beetle outbreak in British Columbia. The pine beetle is an insect that has a special talent for turning a forest into firewood. Previously, pine beetles didn't survive the winter. Due to warmer winters, the insect numbers hit catastrophic levels and wiped out an area of forest twice the size of Sweden. All that standing dead wood combined with rising temperatures to create a perfect storm for wildfires. Fires in 2017 and 2021 made breathing unsafe over a large area due to smoke, and very dramatically on July 1, 2021, in 15 minutes, burned the entire town of Lytton to the ground before the fire truck could get out of the station. (See Chapter 8 to get better acquainted with pine beetles.)

In fact, because of both fire and increased insect damage, forests in Canada ceased to be a net sink (refer to Chapter 2) for carbon in the mid-1970s. Canada's forests still hold millions of tons of carbon, but on an annual basis, these forests now give off more carbon than they suck in. Adding to this vicious cycle, forest fires pump carbon dioxide into the air when the wood burns and releases the gas.

Recognizing how they start and how to prevent them

When trees are tinder dry, it doesn't take much to start a wildfire. Even dry grass can cause devastating fires as happened near Denver, Colorado, in late 2021. Successive years of drought create fuel for fires. When local practices don't remove that fuel, the chances of fires increase. Communities and larger governments need to promote *fire smarting* to remove the fuel on the forest floor. Controlled burning as was practiced by indigenous people can help keep wildfires from burning too long or too hot and out of control.

When conditions are perfect for fires, it doesn't take much to cause a disaster. A carelessly tossed cigarette but, a spark from heavy equipment, and quite commonly a lightning strike can cause a massive fire. In the summer of 2021, with more than 1,600 significant forest fires across British Columbia, we learned a new word *pyrocumulonimbus* — clouds that are formed by fire and rise very high into the atmosphere. Often they carry burning materials from the original fire that can fall to earth far and ignite new fires.

Turning Up the Heat

You may think that we're just stating the obvious when we say that global warming will bring about more hot days and warm nights. But those hot days can be fatal, particularly when they constitute a *heat wave,* a prolonged period of very hot weather. Scientists added a new term in 2021 as Earth experienced heat domes for the first time. A *heat dome* occurs when the hot ocean air traps the atmosphere like a cap or a lid. The 2021 heat dome in British Columbia, where temperatures hit nearly 50 degrees C, killed more than 600 people in four days. That same heat dome extended through much of the United States Pacific Northwest.

In fact, in both the United States and Europe, heatwaves kill more people each year than tornadoes, floods, and hurricanes combined. An estimated 35,000 people died in Europe because of extreme heatwaves in the summer of 2003. During a five-day heatwave in Chicago in July 1995, several hundred died. You don't necessarily hear about other heatwaves: Many countries across Africa have been enduring heatwaves that last for longer periods of time than they have in the past.

High temperatures can mean high stress on the body, particularly heatstroke. Heatstroke, in extreme cases, can lead to chronic illness and sometimes death (see Chapter 9 for more on global warming's effects on people). Society's most vulnerable — the poor, the elderly, and (especially) the elderly poor are usually the victims of killer heat.

Heatwaves also claim livestock. Heat stress can lower livestock's ability to reproduce and increase mortality rates. News reports show that the 2006 heat wave that hit California killed 25,000 cattle and 700,000 chickens and turkeys. The 2021 heat dome in 2021 that killed more than 600 people also killed 650,000 animals on farms in British Columbia. Meanwhile, that same 2021 heat dome killed vast numbers of birds in their nests and wild animals on sea and on land. Scientists estimated that more than one billion sea creatures died on the B.C. coast.

And when people want to cool off, they're adding to another problem — air conditioning and refrigerators working overtime commonly raise the amount of electricity used in cities during heatwaves. Climate change will raise peak energy demands because people will reach for the air-conditioning dial, making conservation efforts more difficult.

Examining the Negative Side Effects of Positive Feedback Loops

The carbon cycle, which we talk about in Chapter 2, is like Earth's respiratory system. Some organisms emit carbon, and others breathe it in. When the carbon cycle is balanced, it's a natural wonder, ensuring that the air doesn't contain an excess of carbon dioxide.

When the carbon cycle goes off-kilter because of too many carbon emitters and too few carbon absorbers, it could prove disastrous. The imbalanced carbon cycle is creating positive feedback loops. A positive feedback loop is anything but good news, even if it does have the word "positive" in its name. In a *positive feedback loop*, effects are perpetually amplified. In the case of the carbon cycle, constantly escalating carbon emissions cause ever-increasing temperatures. If the feedback loops accelerate, there's a possibility of a *run-away greenhouse effect.*

Here's an example of the run-away greenhouse effect in action: A high amount of human-produced carbon dioxide in the atmosphere intensifies the greenhouse effect, leading to higher temperatures that melt the once permanently frozen ground (known as *permafrost*) in the Arctic. That frozen ground has been a GHG reservoir for thousands of years, storing methane that it releases into the atmosphere when it melts. And that additional methane heats things up even more, which causes more carbon emissions. (You can see how this situation gets into a worsening loop.)

A DIM HOPE TO FIGHT GLOBAL WARMING

When a great deal of particulate matter — such as volcanic ash, dust, pollution, and gas from aerosols — prevents some of the sun's light from reaching Earth, it's referred to as *global dimming.* For a while, scientists hoped that global dimming might help correct global warming — less sunlight reaching the Earth would cool things down considerably. But NASA reported in 2006 that the amount of particulate matter in the atmosphere is decreasing — which could be due to the effect of stricter air quality regulations on industry — so more of the sun's light reaches the planet.

The same kind of vicious cycle happens when Arctic ice melts. When the ice is present, the sun strikes the white surface and bounces back, just like a mirror reflecting light. You know the handy trick for dressing on hot days? White clothing keeps you cooler because white reflects the sun's rays, and black clothing soaks in the heat. Well, Arctic scientists call this the *albedo effect*.

When the ice melts, dark ocean water replaces it. The dark ocean water doesn't reflect the sunlight. (Just like your black T-shirt doesn't keep you cool.) It soaks up the heat, warming ocean water even faster and melting the ice more quickly, which leads to more open water and more warming. Warmer waters might also mean that other organisms can thrive where they haven't before, pushing out algae and lowering the amount of carbon dioxide being sucked up. Figure 7-5 shows the sizeable contrast between the white ice and the comparably black water — you can imagine how much more heat the water absorbs than the ice does.

Positive feedback loops also apply when forests catch fire because of warm and dry conditions. Those burning forests release the stored carbon in their branches, trunks, and leaves, adding to the carbon dioxide in the atmosphere. That additional carbon dioxide leads to more warming and dryer conditions, which lead to more forest fires and wildfires.

See Figure 7-6 for a visual representation of other positive feedback loops.

WARNING

Some scientists worry that if humanity doesn't dramatically reduce GHG emissions, the positive feedback loops could take over. If the positive feedback loops take control, reservoirs of carbon that represent thousands of years' worth of carbon sucked from the atmosphere would rapidly release their GHGs. A run-away greenhouse effect — sometimes called "Hothouse Earth" — could, theoretically, end life on Earth. The IPCC calls this a "low probability" event (refer to the nearby sidebar around probability events). Humanity has only a short time to avoid global average temperature rise to more than 3.6 degrees Fahrenheit (2 degrees Celsius) above 1850 at which point the risks increase.

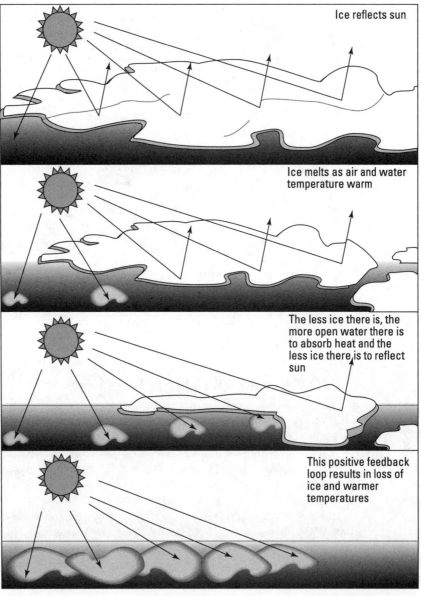

Ice reflects sun

Ice melts as air and water temperature warm

The less ice there is, the more open water there is to absorb heat and the less ice there is to reflect sun

This positive feedback loop results in loss of ice and warmer temperatures

FIGURE 7-5:
White ice reflects heat, but dark water absorbs it.

© *John Wiley & Sons, Inc.*

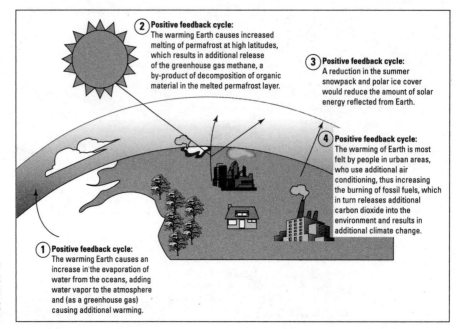

FIGURE 7-6:
Positive feedback loops that global warming might cause.

Inside figure:

2 Positive feedback cycle: The warming Earth causes increased melting of permafrost at high latitudes, which results in additional release of the greenhouse gas methane, a by-product of decomposition of organic material in the melted permafrost layer.

3 Positive feedback cycle: A reduction in the summer snowpack and polar ice cover would reduce the amount of solar energy reflected from Earth.

4 Positive feedback cycle: The warming of Earth is most felt by people in urban areas, who use additional air conditioning, thus increasing the burning of fossil fuels, which in turn releases additional carbon dioxide into the environment and results in additional climate change.

1 Positive feedback cycle: The warming Earth causes an increase in the evaporation of water from the oceans, adding water vapor to the atmosphere and (as a greenhouse gas) causing additional warming.

NOT SO PERMANENT

Unfortunately, permafrost is already thawing in the western Arctic of Canada and Russia's Siberia. When permafrost thaws, it releases into the atmosphere methane gas that had been stored for thousands of years. The melting alone is bad news for local people. Roads collapse when the ground subsides. Houses and villages have had to be relocated.

The Inuit — indigenous people of the circumpolar region, known largely as Inupiat Eskimo in Alaska and Inuit in Canada and Greenland — used to put foods that needed refrigeration into the permafrost, but they can no longer do that because the ground is warming up and the permafrost is disappearing. Their natural fridge is defrosting.

Chapter 8

Risking Flora and Fauna: Impacts on Plants and Animals

S ome people call it nature. Some talk about the birds and the bees, all crea-tures great and small. Others talk about flora and fauna. Scientists call the planet's variety of living species, from the genetic level to the landscape level, *biodiversity* or *biological diversity*. The living world is amazingly diverse, with all sorts of plants, animals, insects, fungi, bacteria, and so on, from the deep for-ests of China to the mountains of Canada to the icy waters of Antarctica.

The more diverse a living system is, the more likely it will be resilient and adapt-able to changes. Biodiversity is affected by many issues — the impact of people, the way species interact with each other, changes in the environment around communities, and so on. But of all the impacts on the health of the living bio-sphere, climate change is now the most destructive.

The United Nations Convention for the Protection of Biological Diversity first identified climate change as a threat to biodiversity. While the natural world becomes increasingly threatened, the way humans use and interact with plants and animals will change. More or less, all the living things on the planet are in this together. In this chapter, we look at what kind of pressure climate change is putting on the Earth's biodiversity, whether organisms can adapt to these changes, and which plants and animals will likely be most affected.

Understanding the Stresses on Ecosystems

If you've ever watched a nature program on television, you know that the natural world isn't the most peaceful of places, with predators mercilessly chasing down prey. But if you've been out on a hike, nature may appear to actually be a pretty harmonious, finely balanced arrangement. The relationships of living creatures with each other, the land, and the climate within a particular area are known as an *ecosystem.*

These sections focus on how ecosystems respond to climate change and how limitations keep species from adapting.

Ecosystems can adapt to some climate change

Ecosystems can be highly adaptable. For hundreds of millions of years, the Earth's ecosystems have adapted to changing climates, responding to differences in rainfall, shifting temperatures, available land, and even changes in the levels of carbon dioxide in the air. But now, climate change is occurring at such a rapid pace that many species likely won't be able to adapt, and entire ecosystems may be transformed.

Worse still, global warming is happening in combination with many other human-created pressures on ecosystems. With 7.8 billion people on the planet, and on the way to 10.9 billion people by 2100, ecosystems are under much greater pressure than they ever were before. If you've ever looked out the window of a plane while flying over land, you've seen human-made patches and lines across the land — patches for developments such as cities, farming, and logging; and lines for structures such as highways, roads, and railroad tracks. Every patch or line that people make fragments the natural ecosystems. Often, people conserve a patch of forest, but human development surrounds that forest, so the plants and animals must rely on each other in that reduced area, allowing less room for adaptation. Air and water pollution hurt biodiversity, too. Humanity has driven many species to extinction.

Faced with a change in their environment (for example, warmer temperatures), species often adapt by moving to a climate to which they're better suited, but plants and animals that live on small islands, on mountain peaks, or along coastlines often don't have this option. Species living in areas that have been fragmented by human developments — forests surrounded by subdivisions, for example — also can't easily move.

Some species can't move very far, no matter what. Slow and steady won't win the race for survival if snails and turtles have to follow their ideal temperature ranges toward the Earth's poles. Think about flightless birds and insects — they're in for a very long walk on very small legs. And most plants stick to their roots (although some plants do manage to travel — see the sidebar "But plants can't move!" in this chapter).

Some species have limitations for adapting

Some plants and animals can live only in certain temperature ranges. For instance, Alpine meadows are very dependent on cool temperatures along the tops of mountain ranges. Coral reefs can bleach out and die because of a very small temperature change in the surrounding ocean. In Australia, some species of eucalyptus trees can survive no more than a 1.8 degree Fahrenheit (1 degree Celsius) temperature shift.

REMEMBER

The Intergovernmental Panel on Climate Change (IPCC) says that an average increase in global warming of 3.6 degrees Fahrenheit (2 degrees Celsius) above 1850 levels — or 2.2 degrees Fahrenheit (1.2 degrees Celsius) above today's temperatures — will have serious effects on all the world's major ecosystems, both on land and in water. (This shift in temperature is sometimes called the *danger zone*, which we discuss in more detail in Chapter 3.)

Ecosystems typically take from a few years to a few centuries to adapt to climate changes — which means that some ecosystems may not be able to cope in their current form. Even if alligators used to swim in the swamps of the current-day Arctic, never before have changes happened at such rapid rates and in combination with so many other pressures that human society has placed on ecosystems. The UN Convention for the Protection of Biological Diversity specifically states that "climate change is likely to become the dominant direct driver of biodiversity loss by the end of the century."

Scientists don't fully understand how ecosystems will be affected by continuing climate change. Ecologists don't know how organisms will respond to the stresses on their ecosystems. Climate change will affect each and every species in different ways. Some species may need to move, others may need to start eating different foods.

BUT PLANTS CAN'T MOVE!

We know, we know — we keep talking about plants and animals adapting by moving elsewhere. But how does a plant or tree move? Of course, individual plants can't travel, but the seeds that the plants spread can. Here's how:

- **Poop:** Trees or plants that grow fruit or berries (think apple trees or raspberry bushes) are a food source for animals and people. When you eat a raspberry and swallow the seeds, well . . . what goes in one end must come out the other! Animals in the wild eat the fruit, wander while they digest, then do their business. The seeds in the poop germinate in the ground and grow a raspberry bush.

- **Picked up, dropped down:** Some fruits, such as plums and peaches, have big pits. Birds and animals take the fruit, eat it somewhere else, and discard the pit — which potentially may blossom into a new tree.

- **Wind:** Plants produce pollen and seeds that, after they develop, get carried away by the wind. Sometimes, they don't go far, but they can! For example, a tree drops pine cones, and those pine cones can also be spread by the wind.

GOOD NEWS

Healthy ecosystems can help humanity adapt to climate change and even reduce greenhouse gas (GHG) emissions. The negotiators at COP26 in Glasgow focused on *nature-based climate solutions* (NBCS). As we discuss in Chapter 2, plants and healthy soils take in carbon dioxide. Plants create a lower temperature, and their root systems keep the soil healthy and able to absorb water, instead of letting it run off. The Convention on Biological Diversity reports that healthy and diverse ecosystems are more likely to adapt to climate change than are those that aren't healthy or diverse.

Warming the World's Waters: Threats to the Underwater World

Water makes up 70 percent of the Earth's surface, making it a very important set of ecosystems, including oceans, seas, wetlands, rivers, streams, and swamps. Climate change will affect all of these ecosystems in the form of increasing water temperatures, rising sea levels, and intense rain or equally intense droughts. (Refer to Chapter 7 for more about the natural disasters that global warming may cause.) Exactly how these ecosystems will be affected, no one knows. Climate change is reshuffling the deck of water systems, and the world doesn't know what kind of hand it'll get dealt.

Many fish species are already at risk of extinction due to overfishing. According to scientists at Dalhousie University in Canada, the ocean's fish populations have dropped more than 30 percent since the 1950s and are continuing to decrease. The projected loss from climate change will further elevate the risk of extinction. Meanwhile, the number of migratory freshwater fish has dropped 76 percent in the 50 years between 1970 and 2016, per *National Geographic*.

Whether a few species are thriving or many species are declining in an ecosystem, these differences change the way the ecosystem functions. Each organism has a role. Ecosystems are remarkably adaptable; change the role of one organism, and the whole system alters in response. Global warming's impacts on ecosystems may cause some radical changes in their composition and how they function.

The next sections take the temperature (figuratively!) of how climate impacts hit the natural world — under ocean water, in rivers and streams, in forests, and at the poles, Arctic and Antarctic. No place on this planet will be spared at least some serious impacts.

Under the sea

Ocean ecosystems are very complex. Climate change is already changing these ecosystems dramatically, proving disastrous for some species. The World Conservation Monitoring Centre highlights pressures brought on by global warming that they project will particularly affect ocean life:

>> **Increasing carbon dioxide in the water:** When more carbon dioxide is in the atmosphere, some dissolves in the ocean water and creates *carbonic acid,* and the water becomes more acidic. Rising acidity makes it more difficult for sea snails or coral toto build their shells. When those species are endangered, so are all the species that prey on them or in the case of coral reefs, those that live in and depend on them.

>> **Increasing water temperature:** Many ocean species are sensitive to the water temperature. Coral reefs, for example, have been shown to suffer badly from higher water temperatures. The Great Barrier Reef off the coast of Australia is at extreme risk. We discuss coral reefs more in the section, "Coral at risk," later in this chapter.

>> **Shifting ocean movements:** The ocean is constantly in motion, with vast conveyor belts of currents that carry warmth to cooler parts of the globe (and vice versa). These currents provide food sources for other sea creatures. Increased fresh meltwater from polar ice, brought on by climate change, has the potential to shift, stall, or stop some ocean currents altogether. (Refer to Chapter 7 for more information on the potential impact on ocean currents.) Currents influence the heat transfer in ocean environments, and changes in how warm water moves around can have consequences for temperature-sensitive species.

The impact of these changes will be different throughout the oceans of the world, which we discuss in the following sections.

Drops in productivity

The IPCC expects icy ocean ecosystems in the Arctic and Antarctic to drop in productivity by 42 percent and 17 percent, respectively, by the year 2050. A productive ecosystem is one that's conducive to life; when an ecosystem's productivity drops, organisms within it decline. Much of this expected drop in productivity in the polar regions of the world relates to the expected decline in phytoplankton. *Phytoplankton* is a microscopic plant that's called algae when it's clumped together (but it isn't the same thing as blue-green algae, which are bacteria), and it's the primary food source for the entire ocean food web.

Phytoplankton in ice ecosystems forms along the edge of the sea ice. When sea ice shrinks, so does its edge, limiting the area that produces phytoplankton. This reduction in phytoplankton will affect the fish and krill that feed on phytoplankton, their predators (penguins), and *their* predators (seals). See Figure 8-1 for a simplified version of the ocean food chain.

REMEMBER

Research from the Institute of Science in Society notes that, without phytoplankton, "marine life will literally starve to death."

Coral at risk

Coral reefs are already among the marine ecosystems at greatest risk of being destroyed by the effects of climate change. They're the hub of diversity in the oceans. Coral might look like a rock or a rocky plant, but what you're seeing is actually the outer skeleton of a living animal — yes, an animal — that has a stomach, a mouth, and even a sex life. Corals come in all shapes and sizes, twisting from left to right and in a rainbow of colors. They live all over the world, in both warm and cold waters, but they're mostly concentrated around Southeast Asia, the Caribbean, and Australia. The reefs support a wide array of small living water plants and fish. Without the reef, the plants and fish have no home. And without the plants and fish, the bigger fish in the sea have no food, and so on.

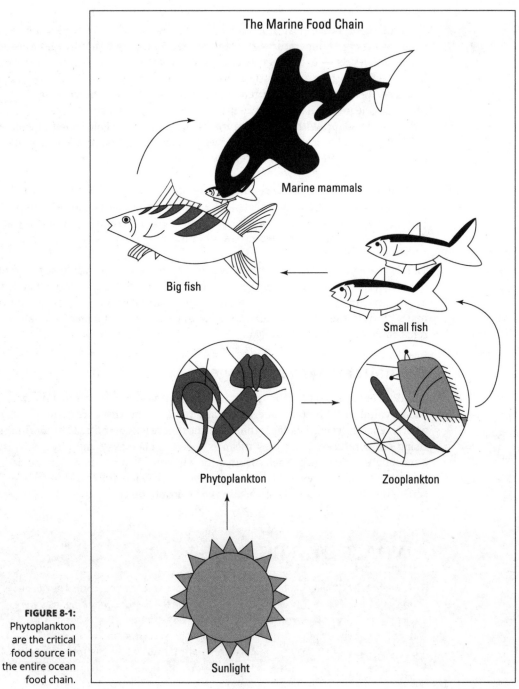

The Marine Food Chain

Marine mammals

Big fish

Small fish

Phytoplankton

Zooplankton

Sunlight

FIGURE 8-1:
Phytoplankton
are the critical
food source in
the entire ocean
food chain.

© John Wiley & Sons, Inc.

The IPCC reports that most, if not all, coral reefs will be bleached (meaning dead) if global average temperatures rise 3.1 degrees Fahrenheit (1.7 degrees Celsius) above 1850 levels — or 1.6 degrees Fahrenheit (0.9 degrees Celsius) above current levels. A sea surface temperature rise of 3.6 to 5.4 degrees Fahrenheit (2 to 3 degrees Celsius) above 1850 temperatures — or 0.4 to 4 degrees Fahrenheit (0.2 to 2.2 degrees Celsius) above current levels — is the danger zone for coral, unless it can adapt. With a temperature rise of 5.4 degrees Fahrenheit (3 degrees Celsius) above 1850 levels — or 4 degrees Fahrenheit (2.2 degrees Celsius) above today's levels — all coral reef systems would die.

Despite the increasing evidence linking climate change to coral reef destruction, you can't easily separate the effects of the climate from other major human pressures, such as pollution and fishing. The Caribbean, for example, has already lost 80 percent of its coral reefs because of other human pressures.

The IPCC predicts that areas covered by cold-water corals, such as those found off the east coast of Canada and the west coast of Norway, will go through a huge drop in productivity by the end of the century. Cold-water corals depend on nutrients that sink from the surface or arrive via ocean currents, so changes in ocean currents could threaten their survival.

Dramatic ecosystem changes

The IPCC expects a major change in the way marine ecosystems function worldwide, even if fossil fuel use is dramatically reduced and trees are planted everywhere. What the scientists say is that the atmosphere is "committed" to continuing rising temperatures. What they mean is that the chemistry and physics of the atmosphere are nonnegotiable. Humanity has stoked the furnace for increased warming. It can't be reversed. These changes are now inevitable due to the rise in carbon dioxide levels people already in the atmosphere.

A WHALE OF A CHANGE

Big changes have already happened making big changes around the world. The Atlantic Coast's endangered right whale used to be at home in the Bay of Fundy. Scientists were scratching their heads when the whales left home. Gone from the Bay of Fundy, they had moved to the Gulf of St. Lawrence. The speculation was that the whales followed their food — fish that found the temperatures in the Bay too warm. But the Gulf of St. Lawrence is a dangerous place for whales, with more boats and fishing nets and noise. Not a good place for this endangered species to call home.

While some species decline and others migrate, the animals and plants in the sea will form new relationships. Predators may find themselves prey, and other species may wind up competing with one another for the same food resources. No one knows how easy these transitions will be, nor whether the emissions of GHG will be reduced, and ultimately stopped, at levels that keep global temperatures at a level allowing species to survive to make the transition.

Lakes, rivers, wetlands, and bogs

The world's freshwater will also see some serious changes brought on by global warming. Earlier spring runoff from snow and glacier melt, and less runoff in late summer, will affect rivers in many parts of the world. These rivers will be at higher levels in the spring and at lower levels in the late summer. And the lower summer levels mean less freshwater availability when it is most biologically needed. (Refer to Chapter 7 for more about melting glaciers.)

Wetlands and bogs are also at risk. These ecosystems depend on having enough water, and they don't have much room for adaptation. When temperatures get warmer, more water evaporates from these areas. A dry bog or wetland is a bog or wetland no more.

In the next sections, we explain how warming of lakes and rivers is bad for their health. It impacts water for people to use, and it really hurts the fish! Healthy water has to have a healthy temperature.

Declining water quality

When water temperatures in lakes and rivers go up, the water quality declines. Here's an example of one process that leads to reduced water quality:

1. **Warm waters are good homes for *algae* — tiny, green, plant-like organisms that look like scum on water.**

2. **When the algae fall to the bottom of the waterbed and decompose, underwater sediment gives out phosphorus.**

   Phosphorous is a poisonous chemical that looks like yellow wax and glows in the dark in natural environments. It's also used for man-made products, such as the red bit on the tips of matches. Phosphorous can give a bad taste and odor to drinking water.

3. **Algae thrive on phosphorous.**

   Excess phosphorous means more algae.

4. Algae grow quickly, called an *algal bloom.*

Bacteria and fungi decompose the algae and hog all the oxygen in the water because they breathe it in. The oxygen level in the water falls, meaning the water contains less oxygen for fish — which need oxygen to survive, just like we do.

Decreasing fish species

Scientists expect that warmer temperatures in lakes and rivers will have an adverse effect on fish species. Although some species may thrive — indeed, some fish species populations are actually increasing — others potentially face extinction. Different species have different ranges of temperatures in which they're comfortable. Species that have large comfortable temperature ranges can adapt more easily than more sensitive species because they're not affected by these temperature shifts.

As the saying goes, "you win some, you lose some." Unfortunately, climate change means the Earth will lose a lot. For example, the IPCC predicts that if global temperatures rise 2.3 to 3.1 degrees Fahrenheit (1.3 to 1.7 degrees Celsius) above 1850 levels — or 0.9 to 2.2 degrees Fahrenheit (0.5 to 0.9 degrees Celsius) above current levels — then North America will lose 8 to 16 percent of its freshwater fish habitat. This loss of habitat translates to a 9- to 18-percent loss of salmon. And West Coast salmon are already in crisis.

Risking Earth's Forests

At least at first, warming temperatures were expected to be good for plants and trees. The growing seasons increased while more carbon dioxide in the air gave plants and trees more to take in.

But over the recent past, rises in temperatures and changes in ecosystems made the overall impact on the forests negative. Forests now experience larger, more frequent fires; regular disease outbreaks; and insect infestations, to say nothing of the ever-increasing human deforestation practices worldwide.

Forests are also hurt by rising temperatures. Whole forests, like the Amazon, could disappear. In these next sections, you get the lowdown on the die down.

Tropical

Rainforests, which are most commonly found in the tropics, can survive only within small temperature ranges. As their name suggests, rainforests depend on rain. The combination of warming temperatures and changing rainfall patterns brought on by global warming could adversely affect the rainforests. In that situation, the IPCC predicts that species in tropical mountain forests face a high risk of extinction. IPCC reports show that the Amazon region could lose a lot of its forest and biodiversity if global average temperatures reach 4.5 degrees Fahrenheit (2.5 degrees Celsius) above 1850 levels — or 3.1 degrees Fahrenheit (1.7 degrees Celsius) above current levels. In fact, if the concentrations of carbon dioxide were to grow by 50 percent, the scientists at the Meteorological Office Hadley Centre in the U.K. say the Amazon rainforest could disappear.

Initially, scientists had thought that temperature changes wouldn't fluctuate much in regions around the equator in comparison to northern regions. Climate scientists were shocked when studies showed the Brazilian Amazon region had temperature increases on the same order as those seen in the Arctic — but only two months of the year. The Amazon Rainforest, once the great carbon sink for the planet, is now a net emitter — it releases more carbon dioxide than it takes in.

GOOD
NEWS

Scientists increasingly see forest planting — and lots of it — as an important climate solution. One forest type that's very easy to restore is the mangrove forests in the equatorial zone. And because they root in the water, they aren't likely to catch fire and set back climate efforts.

Boreal

Boreal forests bring to mind the famous Aurora borealis (the "Northern Lights"), but the name *boreal* simply means northern. These vast coniferous forests ring the northern region of the globe, from Scandinavia (where it's called *taiga*) to Russia, Alaska, and much of Canada. Boreal forests are found in northern and high Alpine regions — think pine, spruce, and fir trees.

Boreal forests will undergo many small-scale shifts while the climate changes, including the following:

>> Changes in the numbers of species living in the forest

>> Faster maturation and shorter life spans for trees

>> Shifting relationships within the ecosystem

Rainfall patterns are also expected to change, and in many cases decrease, bringing more droughts. Taken together, these changes will add up to a major transformation in temperate forests.

Boreal forests have always been known to be a fire-driven ecosystem. In other words, the very ecology of these vast northern coniferous forests required large-scale fires from time to time. But now due to fires and increased insect assault, the boreal forest has ceased to be a carbon sink. It now releases more carbon than it absorbs.

WARNING

The forest industry has reason to worry. What is called *forest biomass* — the amount of forest fiber to harvest — will be reduced. There will also be a decline in wildlife in the boreal.

As you read in the next few sections, climate impacts on forests around the world harm the forests while also undermining the natural role of forests in keeping carbon out of the atmosphere.

More forest fires

Fires are a normal, dynamic part of the way the ecosystem currently functions in boreal forests. Fires clear out older trees and enable saplings to flourish. Such fires haven't been a regular occurrence, however, and they haven't spread like . . . well, wildfire. But already increased numbers of fires are affecting larger areas of forest as a result of drier conditions brought on by human-triggered climate change.

Increased pests

Forest pests have already increased as the climate changes. Usually, colder temperatures keep insect populations down — most don't survive the cold winter months. With warming temperatures, however, more insects are making it through the winter. Also, certain kinds of beetles attack old or weak trees — and forests in British Columbia, Canada, used to have an ample supply of old lodgepole pine. Warming temperatures combined with the age structure of the trees led to a devastating insect epidemic and economic disaster for forest-industry dependent communities.

Mountain pine beetles in interior British Columbia have killed millions of trees that were the core of the province's logging economy — an area of forest as large as two Swedens. Lodgepole pine stands, habitat to mountain caribou and an important economic mainstay have been significantly reduced. The large stands of dry, dead trees left behind fueled a large number of wildfires. A similar phenomenon happened in Norway with the spruce bark beetle.

Preparing for Mass Extinctions

When the forces acting on an ecosystem change dramatically, many of the species that live in that ecosystem are at risk.

Although scientists aren't certain how ecosystems will adapt to global warming, and even though each species and region will react differently, the IPCC projects that 18 to 24 percent of plant and animal species will go extinct if the Earth experiences an average global temperature rise of 2.9 to 4.1 degrees Fahrenheit (1.6 to 2.3 degrees Celsius) above 1850 levels — or 1.4 to 2.5 degrees Fahrenheit (0.8 to 1.4 degrees Celsius) above current levels. This extinction estimate keeps increasing as the temperature does, rising to 35 percent extinction with average global temperatures rising 2.9 degrees Fahrenheit (1.6 degrees Celsius) above 1850 levels — or 1.4 degrees Fahrenheit (0.8 degrees Celsius) above current levels.

REMEMBER

Extinctions are irreversible. When those creatures are gone, they're gone!

Overall, species that migrate are expected to be the most vulnerable because their livelihood depends on the climate of the seasons. Migratory species include birds that fly south for the winter and large mammals, such as caribou, that migrate to find food. These animals are at risk because the season in which their food is available may shift more quickly than they can adapt. Species may miss feeding seasons all together. The seasons in the regions between which they migrate may no longer align. For example, red-wing blackbirds might arrive back in their northern homes to find that the marshy areas where they traditionally nest in spring have dried up. The Arctic Climate Impact Assessment by the Arctic Council expects that migratory birds will lose half of their breeding area sometime within this century.

Food sources might increase for some animals because of warmer growing temperatures, but other species, such as many kinds of birds, will suffer, when the area of their habitat declines. Figure 8-2 shows the IPCC's 2007 projections of the different effects that the temperature change will have on wildlife. (A more recent IPCC report on this subject (2014) gave the same data — by the time this book is published the 2022 Sixth Assessment Report should be published, which will have new, and probably more alarming, conclusions.)

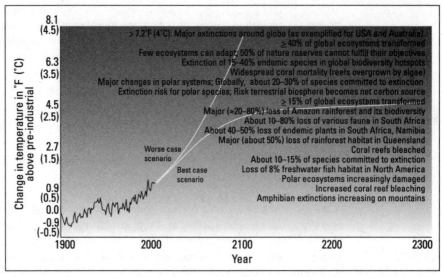

FIGURE 8-2: How climate change could continue to affect ecosystems.

Based on Figure TS.6, Technical Summary. Climate Change 2007: Impacts, Adaptation and Vulnerability. Fourth Assessment Report. *IPCC. Cambridge University Press.*

Life's no beach: Endangered tropical species

Plants and animals in tropical mountain regions are vulnerable to *water stress*, meaning that they either have too much or too little water. Reports from around the world show that warming temperatures affect a number of species in a variety of ways:

» **Small mammals:** Animals that have small populations (and thus a small gene pool) or small habitat ranges are at risk — a handful of these are in Australia, such as possums, bandicoots, and wallaroos.

» **Water birds:** More frequent instances of drought and lowered water tables put birds such as the Baikal teal in Asia at risk because they depend on water areas, such as marshes, to nest and breed.

» **Amphibians and reptiles:** The likes of frogs and lizards could face extinction with the amount of warming that has already happened. Disease outbreaks in some amphibian and reptile populations are occurring because climates are becoming more favorable for those diseases.

Species living in deserts (yes, things can live in the desert!) are at risk because they have to move farther to find a new suitable climate. Areas at high elevations differ greatly from land at sea level in humidity, precipitation, and temperature because of the difference in altitude, so species don't have to look far for a new home. Deserts are flat, however, and pretty much the same across the entire region. For example, 2,800 of the plant species in the Succulent Karoo ecoregion of South Africa will be extinct if temperatures rise 2.7 to 4.9 degrees Fahrenheit (1.5 to 2.7 degrees Celsius) above 1850 temperatures — or 1.3 to 3.4 degrees Fahrenheit (0.7 to 1.9 degrees Celsius) above today's temperatures — because they won't be able to move far and fast enough to find a livable climate. For the same reason, the Jico deer mice and pocket gophers that live in Mexican deserts likely won't be able to adapt to climate change.

Thin ice: Polar bears and other polar animals

Species in polar regions are the most vulnerable in the world to climate change. Warmer temperatures at the North and South Poles will decrease ice cover, increase the temperatures of the water, reduce snow cover, and thaw what used to be permanently frozen ground. The rate of climate change may be too fast for plants to move toward the poles. These changes will put many polar species at risk of extinction, including polar bears, seals, and various birds. Here we explore the two poles in greater detail.

Changes in the Arctic

Because the polar bear is the first large mammal that's facing extinction from global warming, it's become the iconic image for climate change. Scientists estimate that between 22,000 and 31,000 polar bears in the wild are left around the world. Polar bears feed on seals, which spend a lot of their time on the ice. Less ice coverage means fewer hunting opportunities for polar bears. Their lives are entirely dependent on the sea ice.

The melting ice is also bad news for the seals, even if it does reduce their chances of being claimed by a polar bear. All Arctic seals that depend on the ice for resting, breeding, and giving birth — including the bearded, ribbon, and ringed seals — are being put at risk by the ice melt.

In northern Canada, several species will be affected by the shrinking Arctic tundra, which is expected to shrink to a third of its original land cover while temperatures continue to warm. Because they'll be losing their primary source of food — species such as tundra plants — caribou and muskoxen are at risk. The Arctic tundra is also a breeding ground for geese, shorebirds, and the Siberian Crane. When the tundra shrinks, so does the area available for breeding, reducing the chance of breeding, as well.

Extinction in Antarctica

Heading down south to the Antarctic continent — known as the last great wilderness, where humans have seldom been — many species are already at risk:

» Crabeater seals don't depend on crabs to survive — despite the name. They depend on the ice, just like Arctic seals, for resting, breeding, and giving birth. As ice conditions in the Antarctic change, these seals are affected, too.

» Emperor penguins, the movie stars of *March of the Penguins* and *Happy Feet*, eat krill as part of their daily diet. With warmer waters and fewer krill, Emperor penguins have a smaller food source. Warming temperatures make for better breeding because fewer chicks die from extreme cold, but this benefit will be small compared to the loss of krill as a food source, meaning more penguins with less food.

Zoë Caron, one of the authors of the first edition of this book, was lucky enough to stand among thousands of Adélie penguins and their hatching chicks on an Antarctic expedition in 2007. Wildlife specialists on her expedition explained that precipitation is increasing over many parts of the Western Peninsula, where Adélie penguins thrive. So, the peninsula experiences more snow cover more often, even if it melts fairly quickly. These penguins need snow-free surfaces for hatching chicks, so this increased snowfall is a problem for them. Adélies could become extinct in some areas of Antarctica over the next 7 to 30 years.

Chapter **9**

Hitting Home: Global Warming's Direct Effect on People

W hen this book first came out in 2009, we were able to say, "So far, global warming hasn't had a huge impact on most people's lives."

Unfortunately, that's increasingly not the case. Whether it's a farmer dealing with drought, a New Yorker trying to wade through the subway, a European dealing with flash floods, or an Indian coping with massive heatwaves, that statement is increasingly untrue.

The impact of global warming will increase in the coming years, but the degree of change will still vary greatly, depending on a few factors, including the following:

>> **A person's location:** Where you live matters. Someone living 1 meter above sea level will experience a much larger impact than someone at a higher elevation. Generally speaking, living closer to the poles will have a more extreme change than at mid-latitudes. And living in a wealthy country with resources to help in emergencies is extremely relevant.

>> **How rapidly nations around the world reduce greenhouse emissions:** No matter where you live, the unchecked impacts of climate change are potentially catastrophic in the long term. The faster humanity stops using fossil fuels, the safer everyone will be.

>> **A person's wealth:** And just as rich countries are better off, people with the resources to shield themselves can take steps to reduce the immediate, short-term costs and stress levels. People without those resources won't be so lucky. Those with money will be able to pay their way out of problems at first — whether it means building new infrastructures to protect people from natural disasters or paying for importing water in the instance of drought. But not enough money in the world will protect the wealthy from levels of climate crisis that make the planet increasingly uninhabitable (perhaps that's why some of the billionaire class are so eager to go to Mars). Unless humanity acts fast, both the rich and the poor will suffer eventually. Everyone is in the same boat.

In this chapter, we look at how changes in climate can affect your health, food, and city. We also consider how some people — such as northern indigenous communities and women and children around the world — will be particularly adversely affected by global warming.

Like all the information in this Part, this chapter may get your spirits down. But remember, these projections merely suggest what might happen if civilization stays its current course. People can stop the impact of global warming from becoming catastrophic — and some effects can be avoided entirely — but more people and their governments must start cutting carbon emissions right now. Check out Parts 4 and 5 for more about how civilization can avoid the worst-case scenarios that we discuss in this chapter.

Focusing on the Health Scare — Outbreaks and Diseases

In 2009 the World Health Organization (WHO) estimated that climate change was responsible for 150,000 individual deaths around the world in the year 2000. Now the WHO is predicting 250,000 deaths a year from 2030 to 2050. Climate change is already affecting people's health.

The bad impacts on health will, unfortunately, greatly outweigh the good. Climate change is bad for human health in several major ways:

>> People are killed or injured in extreme weather events made more frequent and intense by climate crisis.

>> Extreme events make it harder for people to survive — such as when prolonged drought endangers food security, driving people from their homes, becoming internally displaced in their own country, or becoming refugees.

>> People become sick from climate change–related events and die of other illnesses. Millions of people around the world die from diseases caused by air pollution from burning fossil fuels. Others die from disease carried by insects (known as *vector-borne*) diseases, like dengue fever or Lyme disease. WHO has stated that climate change is the single biggest health threat to humanity.

As we discuss in Chapter 7, heatwaves and more severe heat domes are killers — the impact of heat stroke is serious as a result. At local and regional levels, first responders and health authorities need to be better prepared to save lives. For example, before wildfires strike, individuals and communities can take steps to reduce the impact of fires by *fire smarting*, which includes identifying wildfire hazards, making smart choices of plants, shrubs, grass, and mulch, and taking necessary precautions to protect homes and buildings.

REMEMBER

Climate adaptation and resilience are key, and everyone has a role to play. Even simple steps like being sure to check on elderly neighbors and encouraging them to move to an air-conditioned space — even a store around the corner. For first responders, better protocols are needed. Putting over-heated people in ambulances may be a fatal mistake. Saving lives can mean putting them fully clothed in a cool bath.

Some parts of the planet are already showing early signs that they'll become uninhabitable — too hot to live there even with air conditioning. For example, Delhi the region in India with more than 26 million people is one location where the question is being asked. In the 2015 heatwave, temperatures exceeded 114 degrees Fahrenheit (46 degrees Celsius) — and the humidity made it worse.

In the following sections we discuss different diseases and health conditions that will be accelerated with climate change.

Dengue fever

Found in warm climates, dengue fever carried by mosquitoes causes severe joint pain. A pool of still water is a mosquito's honeymoon suite. Consequently, a lot of rain (creating more pools of water where mosquitoes breed) combined with temperatures warm enough for the mosquitoes to survive can increase its spread — but oddly, so can drought. When drought hits an area, more people store water outdoors, creating other excellent mosquito-breeding sites. The Intergovernmental Panel on Climate Change (IPCC) expects that global warming will create climates in both New Zealand and Australia that are favorable for mosquitoes and for dengue, and that both countries will see more species of mosquitoes that can carry the virus.

One report shows that about a third of the world already has favorable conditions for dengue and that 5 to 6 billion people will be at risk of dengue by 2085, compared to the 3.5 billion that would be at risk without climate change as a factor. Dengue virus infections have doubled every decade since 1990 — with global warming a major factor. A study published in the *Lancet* journal reports that like dengue, other arboviruses like chikungunya and Zika are also increasing with warmer temperatures.

Lyme disease

Due to climate change, according to the Centers for Disease Control (CDC) in Atlanta, Lyme disease has become the fastest growing vector-born disease in the United States and Canada. It's spread by ticks carrying the dangerous bacteria *Borrelia burgdorferi*. As areas become warmer, the ticks' range has spread. Originally, only found in areas around New England, Lyme disease (named for Lyme, Connecticut) has spread to every province of Canada, most of the United States, as well as in Europe. When treated immediately with antibiotics, someone bitten by a bacteria-bearing tick can make a full (and quick) recovery. However, Lyme disease is often misdiagnosed. Left untreated, it can result in complete disability and even death.

GOOD
NEWS

Prevention matters! When going out for hikes, tuck your pant legs you're your socks. Check children, partners, and your pets for ticks after you return indoors. Be sure to have tick-removal tweezers handy. You can easily find tick removal kits online. One such site is https://canlyme.com/product/tick-removal-kit/. The other good news is that the medical profession is increasingly recognizing Lyme disease and is prepared to treat it.

Tragically, in Canada, many Lyme patients have to go to the United States for treatment because a doctor familiar with Lyme disease can be difficult to find.

Other diseases and problems worsened by global warming

The IPCC points out a number of health concerns that civilization already deals with that might worsen because of climate change and extreme weather events. This section addresses them in greater detail.

Allergies

If you have allergies to pollen and dust, those allergies could get worse, depending on where you live. If global warming brings you an early spring, that early spring will bring pollen, too — extending your allergy season. Countries such as Canada, Finland, and the Netherlands will probably be most affected by the increase in pollen because they're undergoing larger seasonal changes than more southern countries.

Contaminated drinking water

This could be a risk for areas that see an increase in rainfall and flooding because of global warming. For example, during Hurricane Katrina in the United States, water supplies became contaminated, and many cases of diarrheal diseases appeared, some of which were fatal. (We talk about flooding and water contamination in detail in Chapter 7.)

Cholera

Global warming doesn't directly cause *cholera*, an infectious disease that causes major cramps and diarrhea, but its effects may create an environment in which the disease can flourish. Cholera isn't fatal if treated promptly, but it can be extremely dangerous in parts of the world with shaky public health systems. Cholera usually happens in areas with bad or nonexistent sewer systems, where human excrement ends up mingling with drinking water.

Flooding brought on by climate change might lead to a cholera outbreak in those regions because floodwaters can wash human and animal waste into sources of drinking water. South Asia, for example, could have an increased risk to and toxicity of cholera if local ocean water temperatures rise, reports the IPCC. (Refer to Chapter 7 for more about the risks of flooding.)

Diarrheal disease

Think extreme diarrhea. This disease occurs most frequently in Australia, Peru, Israel, and islands in the Pacific Ocean when the temperatures soar and rainfall patterns change. Bacteria thrive in higher temperatures, and flooding increases the risk of infecting drinking water sources. Areas that have poor sanitation have even greater risk.

Lung problems

Air pollution often worsens health issues, such as asthma. Smog episodes, for example, are more intense during heatwaves — so, you find more lung problems in big cities than anywhere else. Many greenhouse gases (GHGs) — methane and nitrous oxide to name two — just happen to also pollute the air. The women and children in some developing countries have a higher prevalence of lung cancer because of the smoke from fires that they burn to cook their meals — whereas the men aren't as affected because they're out of the house for most of the day. (Refer to the section, "Feeling the Heat First: Unequal Effects," later in this chapter to find out more how climate change disproportionately affects women.")

Wildfires a result of climate crisis have created unsafe air quality in places around the world — from Greece to British Columbia and California. In 2020, the air quality in coastal Victoria, British Columbia, was worse than Beijing, as a result of fires hundreds of miles away in the interior of the province. The COVID pandemic meant that people were urged to only socialize at distance and outdoors. Competing health advisories directed people to go back inside. Talk about colliding crises. Both were bad news for lungs.

Malaria

The link between *malaria* (a disease carried by mosquitoes that causes chills and fever and can lead to death) and climate change is complicated. Years ago, scientists argued that climate change could cause malaria to spread. That seems to be the case as higher temperatures spread to areas that hadn't been malarial zones. But other factors play a big role, including economic and healthcare improvements. For example, malaria transmission will likely decrease in southeastern Africa, but the risk of malaria will increase in industrialized countries including Australia and the United Kingdom.

Skin cancer

Too much direct sun exposure can cause this form of cancer. In areas that experience rising temperatures, people often want to be outside in the sun more and wear less clothing. To add to the problem, GHGs actually cool the upper layers of the atmosphere, which creates perfect conditions for ozone-depleting gases that create holes in the ozone layer. And that means less protection from the sun. An increased risk of skin cancer is particularly worrisome in places with depleted ozone, such as Australia and New Zealand, where, due to proximity to thinning ozone at the South Pole, people already have a high risk to increased UV radiation and dangerous sun exposure.

Vermin

Rats, mice, and other rodents often carry disease. Flooding or heavy rain pushes these little critters out of their burrows and directly into the paths of humans. Low-income countries that are susceptible to flooding, such as many within Central America, are particularly at risk.

Putting Pressure on the Fields

Because agriculture depends on the land and the climate for its products, this industry will feel the impact of global warming more than any other industry, which could cause increased inconsistency and uncertainty for the world's food supply. The temperature changes, severe storms and rains, and altering growing seasons that we discuss in Chapter 7 will affect farmers worldwide while climate change develops. But those changes won't impact all farmers in the same way.

Affecting farmers

Agricultural productivity could increase or decrease, depending on the region as a result of climate change. Northern Europe, for example, should have better growing seasons, but areas around the Mediterranean probably won't do as well.

The megadroughts of the last number of years in the prairie regions of the United States and Canada have created conditions drier than in the "Dirty Thirties" in the Dust Bowl. On the other hand, regions such as southern Africa will suffer from an increasingly warm and dry climate. There, rainfall isn't only becoming less frequent, but also less predictable. These conditions together are actually reducing the length of the planting season in southern Africa, the opposite of what's happening in the Northern Hemisphere. Tropical regions will likely produce far less food from crops, especially rice.

WARNING

In the long term, the IPCC expects that the overall effects of climate change on agriculture would be negative around the world. If people allow temperatures to continue to increase, those temperatures could exceed what many crops can stand, and those crops won't survive. Some crops in dryer or lower-latitude regions suffer from even a small temperature rise. For example, mango and cotton crops in Peru have a shorter growing season when temperatures are higher than normal. Climate change is also a big threat to coffee crops around the world. In fact, coffee may be the most threatened crop due to the climate crisis. It's tempting to tell governments to "wake up and smell the coffee" while you still can.

Temperature aside, extreme weather events, such as heavy rains or droughts, can stress crops by either drowning or dehydrating them, enough to upset any farmer depending on selling those crops. Pest outbreaks can damage crops, and pest outbreaks become more frequent when the climate becomes more favorable to those pests. These outbreaks aren't easy to project because all pests are different and climates differ regionally. To prepare for these possibilities, *agronomists* (scientists focused on crops and soil) are trying to develop crops that are more pest- and drought-resistant.

Hurting the global food supply

Even though more of the world is adequately fed than was the case decades ago, 821 million people are undernourished. The IPCC sees both opportunities and threats in the impacts of the climate crisis on agriculture. Global warming is already putting serious stress on agriculture in terms of water scarcity (drought) and extreme weather events. These pressures will push up the price of many agricultural products.

Meanwhile, the opportunities lie in reducing GHG emission from agriculture. According to the IPCC, somewhere between 21 and 37 percent of GHGs come from agriculture, including the transport, processing production, and packaging of food products, as well as the emissions from farming itself. A major opportunity for farmers to reduce global warming is *regenerative agriculture*, which refers to restoring and revitalizing soils to create healthier growing conditions while soaking up carbon.

Furthermore, nearly one-third of all food produced ends up as waste. Incredible, but true. From 2010 to 2016, that huge level of waste was responsible for 8 to 10 percent of all global GHG emissions. Looks like a win-win solution: helping the poor be healthy and well-nourished and reducing the amount of emissions from the waste can be part of a climate solution.

In 2019, the IPCC said with a high degree of confidence, "Agriculture and the food system are key to global climate change responses. Combining supply-side actions such as efficient production, transport, and processing with demand-side interventions such as modification of food choices, and reduction of food loss and waste, reduces GHG emissions and enhances food system resilience."

The IPCC expects crop production to drop when just a 1.8- to 3.6-degree Fahrenheit (1- to 2-degree Celsius) local temperature shift occurs. This drop in crop production will then directly increase that region's risk of hunger. And globally, with any increase above 5.4 degrees Fahrenheit (3 degrees Celsius), major crops of corn and wheat will be below normal in low latitude countries, such as Brazil and Kenya, whether farmers will have to adapt. This amount of warming will also stress and even kill livestock in semi-arid areas such as inland eastern Australia or dryer regions of Texas. For agriculture in northern countries, such as Canada, the IPCC reports that temperatures are rising more than the world average — the Arctic, for example, has warmed three times as fast as the rest of the world over the last century.

Paying the Price for Global Warming

Debate over global warming really heats up when it comes to money. Because industry contributes a lot of the excess GHGs in the atmosphere, some governments worry that cutting GHGs would have an adverse effect on the economy.

Reducing GHG emissions will have a financial impact. In 2006, Sir Nicholas Stern, former Senior Economist to the World Bank, reviewed the economic impacts of the climate crisis at the request of the U.K. government. His report, "The Stern Review on the Economics of Climate Change," looked at the financial impact that global warming would have on the world's economy. In his update to the 2006 report, in 2009 he realized he had underestimated the benefits to the economy of acting more quickly. In "Key Elements of a Global Deal on Climate Change," he calculated that acting to reduce GHG emissions would cost the world a cut of 2 percent of global *gross domestic product* (also known as GDP, the global measuring stick of economic wealth) annually over the next several decades. India already spends 2 percent of its GDP on adapting to climate change impacts.

Stern's landmark 2006 report also examined the cost to the world if people did nothing and GHG emissions weren't reduced. If humanity doesn't reduce GHG emissions now, the report found it will cost Earth's population five times more than if it does, resulting in a cut of 5 percent of the GDP — every year. And that's one of the better-case scenarios. The worst-case scenarios show that waiting to reduce emissions could cost the world 20 percent of global GDP or more. Failing to reduce GHGs could cost the world economy $7 trillion!

Looking back at his 2006 report in 2009 Stern found that the economic impact of global warming had been significantly underestimated in his study. That was due to the fact that governments have been too slow to act, so the problem keeps getting worse and the costs continue to rise. In Stern's comments, ten years after the original publication of his report, Stern said, "With hindsight, I now realise that I underestimated the risks. I should have been much stronger in what I said in the report about the costs of inaction. I underplayed the dangers."

REMEMBER

Many European countries have benefited with continued GDP growth because they started acting on climate change decades ago. The Stern reports gave a clear message: The longer humanity waits to act on climate change, the more serious the impacts become, and the more it costs humanity to adapt and recover.

GOOD NEWS

In 2020, new research looked at post-COVID economic recovery. Experts found that climate-friendly policies give the economy the biggest boost. Stern and Nobel prize-winning economist Joseph Stiglitz, along with other colleagues, interviewed economists and central bankers from the G-20 nations and more. In fact, they consulted more than 200 experts from 53 countries, and the news was encouraging. Investing in renewable energy, tree-planting and improving the energy efficiency of buildings turned out to give the best — and greenest — bang for the buck!

Also on the other side of the cost — benefit calculation: Recent estimates of the benefits of the transition to renewable energy suggest that the world could experience a $16 trillion net benefit if the current rates of conversion to renewable sources continue. That's because the cost of renewable energy is falling rapidly, whereas the financial (let alone the environmental) costs of fossil fuels continues to climb. Switching to renewables is already less costly than continuing with fossils, and the net benefit can only continue to grow. We talk more about such massive shifts in Chapter 17.

When talking about climate change, sometime people immediately think about the impact on nature — thinking about climate change as an environmental issue. But the impact of climate change is also on economies. It's a big and costly impact on infrastructure like highways and bridges, water treatment plants and electricity grids — the things that make economies work. The next sections look at how climate change threatens those billions of dollars of investment.

Highways, waterworks, and the other stuff humans build

In Chapter 8, we discuss natural systems at risk because of climate change, but human-built systems are at risk, too. Governments call it *infrastructure* — such as the roads that you drive on and the waterworks that take away sewage and deliver *potable* (suitable for drinking) water. It's the stuff that humans have built to make modern life easier. And civilization built all of this stuff for the climate it used to have.

No matter where you are in the world, the costs from extreme weather — major flooding, fires, landslides, and storms — have been increasing for the last three decades. These natural catastrophes, brought on by civilization's unnatural addition of GHGs to the atmosphere, endanger the infrastructure that's the backbone of cities and towns. The repair bill for the following damages won't be cheap:

>> **Buildings:** Storms and flooding can quickly damage unstable buildings. People living in inadequate housing that is easily damaged by strong winds or storms are especially at risk from the extreme effects of climate change.

>> **Electricity demand:** Hotter days mean cooler buildings when people run their air conditioners to the max. Heatwaves come hand-in-hand with skyrocketing electricity demand and, often, major blackouts.

>> **Sewage systems:** Storms and flooding can also cause the sewage systems in cities to overflow.

>> **Transmission lines:** You've likely seen downed telephone poles and power lines after a big storm. Damaged transmission lines could become a more common sight for many parts of the world.

>> **Transportation:** Highways, roads, and railroad lines will all require more frequent maintenance and repair when they're subjected to extreme weather.

Although everyone will be affected by the physical impacts of climate change, some people will be more vulnerable to negative changes. The poor are particularly at risk. Poverty, combined with a lack of social support, was the main cause of heatwave deaths in the Chicago heatwave of 1995. Poverty also has a differential impact not only between the poor and the rich people in places like Chicago where both the super rich and the very poor live in the same city, it's like that globally too – some countries are very poor. Those countries, just like poor people, create a lot less pollution than the very rich countries. For example, poorer nations have been increasingly seriously damaged by all manner of extreme weather events. In March 2019, Cyclone Idai hit the coastal nations of Malawi, Mozambique, and Zimbabwe killing 1,000 people. Climate change makes such storms more dangerous as much of the damage is due to the volume of rainfall. Warmer air holds more moisture than cooler air.

People in low-level coastal cities — accounting for 10 to 23 percent of the world's population — are vulnerable to sea level rise and flooding from storm surges because of their proximity to the ocean. Cities that don't have much green space will be at greater risk, too; without soil and trees that stabilize the ground and absorb water, these concrete cities are more vulnerable to landslides and flooding.

An unfair split: Costs to the industrialized and developing nations

Industrialized countries, such as the United States, Canada, and the United Kingdom have pumped most of the excess GHG emissions into the atmosphere — through industry; their heavy reliance on cars; and their ever-growing, ravenous consumption of electricity and energy.

Even though the largest emitter of GHGs on an annual basis now is the People's Republic of China, the warming impact on the atmosphere from emissions from the U.K., Canada, and the United States are still exerting a larger warming impact than China's. That's because China's increase in emissions is relatively recent and the vast majority of global emissions since 1850 came from the wealthier northern nations.

Developing countries haven't had the amenities that burn fossil fuels and create GHG emissions. The effects of global warming don't care who's to blame, however. Because developing countries have fewer resources and less financial ability to recover from natural disasters, they'll feel the first major impacts of climate change.

Developing countries don't have the economic resources to adapt to climate change and to recover from its worst effects. The impact of Hurricane Katrina in 2005 makes it still the most destructive hurricane in the United States. Katrina happened to one of the wealthiest nations on Earth. Imagine the impact for a nation such as Honduras or Guatemala if Katrina had made landfall there. The IPCC reports that in 1985 and 1999, for example, natural disasters lost about 2.5 percent of the GDP of the world's richest nations, and the poorest nations suffered a loss of 13.4 percent of their combined GDP. If the world's nations fail to meet the Paris Agreement targets, economists forecast global economic losses of 15 percent of GDP by 2085.

Even though Hurricane Katrina remains the single most costly extreme weather event in the United States at $161 billion, recent years have racked up more damage cumulatively. In 2017, the total damage from extreme weather events in the United States broke records at costs exceeding $306 billion.

REMEMBER

The financial costs that come along with extreme weather events — storms, flooding, and droughts — give any country a huge economic hit. Because they lack the industrialized countries' strong, diversified economies, developing nations have far fewer economic resources to bounce back from such a huge disaster — yet they're the most likely to be hit, and hit first, because of their geographical locations. Ninety percent of deaths caused by natural disasters happen in developing countries. You've likely seen stories in the news about these natural-disaster deaths. When this book was first published, large events were becoming more common, such as the 30,000 people killed from flash floods and landslides in Caracas in 1999, or the 15,000 homes damaged by major flooding in Cape Town in 2001. By 2021, the events are larger and more severe, like flooding in South Sudan displacing more than 800,000 people while 200,000 had to move to escape Cyclone Tauktae. In May 2021, that one cyclone hit India, Sri Lanka, and the Maldives. The damage was costly in lives and dollars. The year 2021 was the fourth time in five years time that damages from climate-related disasters exceeded $100 billion.

The loss of life in heatwaves also disproportionately hits poorer nations. An estimated 6,500 people have died due to heatwaves in India since 2010. When this book was first published, scientists were still cautious about attributing the cause of these extreme events to climate change. But the 2021 IPCC report of Working Group 1 ended that notion of uncertainty. IPCC scientists from around the world expressed high confidence that the human impact on climate is a necessary factor behind the number of severe, more frequent and more deadly events of all kinds.

WARNING

Developing nations need to move quickly to adapt — with measures ranging from regulating freshwater use to building levies to planting forests — to cushion the inevitable impacts of climate change. (We discuss how developing nations can adapt to global warming in Chapter 12.) The money that those countries are spending on adaptation was originally earmarked for further development. Consequently, development is slowing while adaptation measures are growing.

GOOD NEWS

As countries around the world begin to apply solutions to climate change, developing countries may get partners to support their adaptation measures. Carbon markets and country-to-country partnerships within the Paris Agreement coupled with strong climate financing amounting to $100 billion per year are two major tools that can help alleviate pressures on developing nations. We talk about these solution-based projects in Chapter 12. Chapter 17 discusses how developing countries may be able to use new renewable energy technologies to empower growth without adding carbon emissions.

Feeling the Heat First: Unequal Effects

Just as climate change affects regions differently, it affects groups of people differently as well. Unfortunately, the impact of climate change will be most keenly felt by those who have few resources to adapt. Activists who seek to address the imbalance of who's causing climate change and who's being affected by it refer to their cause as *climate justice*. In this section, we look at some of the major climate injustices various populations around the world are experiencing — and will experience.

UPHILL BATTLE FOR GLOBAL TOURISM

Increasingly, the tourism sector has realized it's both a victim of global warming and a cause of faster warming. All those exotic locations take massive jets to bring sunseekers to gorgeous resorts fuel the climate emergency. At the same time the climate emergency has huge and negative impacts on tourism. Ski resorts lack snow. Coral reefs are at risk.

At the COP26 conference in Glasgow in 2021, the tourism industry adopted a decade of work to make tourism sustainable. "The Glasgow Declaration: A Commitment to a Decade of Tourism Climate Action" commits to the following:

"We declare our shared commitment to unite all stakeholders in transforming tourism to deliver effective climate action. We support the global commitment to halve emissions by 2030 and reach net zero as soon as possible before 2050. We will consistently align our actions with the latest scientific recommendations, so as to ensure our approach remains consistent with a rise of no more than 1.5 degrees Celsius above pre-industrial levels by 2100."

Northern Hemisphere communities

The north is seeing some of the strongest changes brought on by global warming, and these changes are threatening the way of life for many communities.

Indigenous communities are particularly affected. The permafrost and thick winter ice that served as their road surfaces are melting. Traditional food sources are becoming hard to obtain; the melting ice makes fishing and hunting difficult. Thin ice has caused several hunting accidents.

Climate change is making the weather less predictable than ever before. Many people think about the weather only when they're going somewhere or heading outside for leisure or recreation. But in these indigenous communities, weather helps define their way of life.

Some northern communities think they can fully adapt to all the changes. Others are considering moving to new locations where they can live more easily, and still others are enjoying longer hunting seasons. But all these communities are experiencing noticeable change as we discuss in the following sections, and it's not slowing down any time soon.

Changing culture and lives

The Inuit have been adapting to these changes, but at the cost of their traditional culture. The Inuit are a hunting people, but they now have shorter and shorter hunting seasons because the ice melts earlier and earlier each year. While their environment changes so rapidly, knowledge that was passed down for generations becomes unreliable. In an interview with the *Nunatsiaq News*, Naalak Nappalak, an elder from Kangiqsujuaq, described how he can no longer confidently predict the weather: "Before we knew by looking at the sky whether there would be storms or if it would be calm. Nowadays just when you think you know how the weather will be, they can change in an instant. It's this inconsistency that is most noticeable."

People living outside of the native village of Kotzebue in northwestern Alaska can travel into the town for supplies only when the ice is sufficiently frozen and stable for travel. Warming temperatures mean a longer thawed season, which means people who need supplies or medical attention have longer to wait. The ice also usually serves as a barrier to storm surges, and less freeze-up over the year can lead to erosion and flooding along beachside roads.

In northern Norway and Finland, the indigenous Saami people have been dependent on herding reindeer for thousands of years. With climate change, grazing is available at different times and different places, and sometimes not available at all. The snow is no longer dependable to hold up the weight of the animals, and they're getting stuck and dying out on the tundra. Age-old patterns of interlocked animal and human behaviour are changing. The entire society is at risk.

People living in the indigenous community of Lovozero in Russia have had to deal with climate change in a number of ways. As one local described it, "Bogs and marshes do not freeze immediately, rhythms change, and we have to change our routes of movement, and this means a whole new system of living is under change. Everything has become more difficult."

Affecting the hunt

Some northerners can hunt more easily because of climate change, however, including those who fish for larger whitefish, as well as clamming harvesters and seal hunters (because shorter periods of sea ice give longer water-based hunting periods). Access to wood for fire fuel has improved, too, because it washes up on shore for a longer period of time each year.

REMEMBER

Although northerners may hunt more easily in the short term, climate change threatens the survival of many species (as we talk about in Chapter 8). Animal populations, such as the seals that depend on sea ice, can become easily stressed by global warming, which translates into stress put on humans in the long-run.

In Qaanaaq, Greenland, people describe that just seven or eight years ago, they used to be able to go out onto the ice to hunt as early as October. Now, these hunting grounds sometimes don't freeze up until January. Temperature, coupled with changing sea currents and wind conditions, causes these changes. The places in this area of Greenland are named for their natural and geographic characteristics. Ironically, some of these places no longer fit their names — *Sermiarsussuaq*, which means "the smaller large glacier," used to cover the landscape all the way to the ocean's edge. The glacier no longer exists.

People in poverty

Poorer populations in developing and industrialized countries are extremely vulnerable to the effects of climate change. Poor people don't have the resources to adapt to extreme changes — especially unexpected ones.

Ten percent of the world's population earns less than $2 per day and largely depends on public services, according to Lifewater, a charity focused on ending the global water and sanitation crisis. This improvement is significant. When this book first came out, half of the world earned below $2/day. Poverty has been reduced globally, even though an unacceptably large number of the human family earn less than $1.90/day. In some developing countries, public services may not be able to cope with the aftermath of extreme weather events brought on by global warming. In other countries, public services don't even exist, leaving the poor with few resources to cope. As the IPCC says, "This does not necessarily mean that 'the poor are lost'; they have other coping mechanisms, but climate change might go beyond what traditional coping mechanisms can handle."

Women (and children)

You've likely heard or read about women's rights and women's equality issues. But you probably don't often hear about women's inequality when it comes to global warming. The people working closely on the climate change issues are only now giving it attention.

You may not realize it, but your *gender* (your sex and how it influences the roles you play in society) affects the degree to which global warming may hurt you. In both rich and poor countries, women tend to bear the brunt of the climate change's negative impacts, mostly because they tend to be whole lot poorer. Women make up the majority of the world's farmers. Women have the majority of the world's caregiving — for infants, children, the ill, and the elderly. And women are disproportionately impacted in widowhood, marriage, and divorce. Here we discuss the ways in which climate change is a women's issue. We can also add that it's a children's issue as well. The Paris Agreement includes language about recognizing the critical role of women as those who are impacted and also as those who make a difference to protect the climate.

Women in developing nations

In many developing countries, women are frequently agricultural workers. They're the main family caretakers responsible for producing food for the family. The International Union for the Conservation of Nature (IUCN) reports that women produce 70 to 80 percent of the food in sub-Saharan Africa. Latin America isn't far behind at 65 percent. Scientists predict that these poorest regions are also the ones to be hardest hit by climate change, which we discuss in the section, "Paying the Price for Global Warming," earlier in this chapter. When climate change negatively affects a region, men often migrate to cities to find better paying jobs, leaving the women and children behind to try to survive on the land, which is no longer productive because of climate impacts such as drought.

How natural disasters caused by climate change affect women

Natural disasters, which will increase in the wake of climate change, affect women and men differently. Women, as main caregivers, are more likely to be indoors — particularly in developing countries — when a disaster occurs and won't be able to escape. Even if they do survive, women tend to stay within the community longer afterward to care for their families, thus exposing themselves to deadly diseases.

Although not linked to global warming, the grave impact that natural disasters have on women can be seen in the death toll from the major Asian tsunami that struck at the end of 2005 and hit the province of Aceh in Indonesia, where 75 percent of those who died were women.

When the death toll from natural disasters has significant gender differences, the resulting gender imbalance in the society can have major, long-term negative consequences. The Asian tsunami left the society with a three-to-one ratio of males to females. With so many mothers gone, the area experienced increases in sexual assaults, prostitution, and a lack of education for girls. Research in this area is still in its infancy, but the IPCC has reported that women are more likely than men to suffer from post–traumatic stress disorders after living through a disaster and that men are likely to commit domestic violence against women after natural disasters. This is worrying because climate change is expected to increase the intensity and frequency of storms and extreme weather events around the world.

GOTELIND ALBER: GETTING GENDER ON THE AGENDA

Gotelind Alber, a German physicist who's worked in the energy and climate policy sector for 30 years, is bringing gender issues into the discussion on climate change. Whether working in the fields of science, technology, climate change, or policy development, she has always worked to put gender on the agenda. She co-founded the international network Women for Climate Justice.

Among her long list of projects is the "Climate for Change: Gender equality and climate policy" project. The goal of this European-based initiative is to

- Help balance the participation of women and men in creating climate-related policies.

- Address the fact that major climate change–related sectors (such as energy, transportation, and buildings) are mostly run by males.

This project's aim is to stress the importance of a balanced gender view in decision making. They're working with ten European cities to integrate a gender perspective into their climate change policies. You can find more information and the ongoing results of this project at www.climateforchange.net.

4

Political Progress: Fighting Global Warming Nationally and Internationally

Get a sense of why governments are essential players in the fight against climate change.

Look at what governments can do at every level, from your local mayor to the leaders of your country.

Examine how international leaders are collaborating because climate change is a global matter.

Understand the special needs and opportunities for developing countries.

Figure out the most effective way for you to add your voice.

Chapter **10**

Voting for Your Future: What Governments Can Do

Many people can be cranky about their country's government and cynical about its motives and whether it can really make any positive changes. And no one likes to pay taxes, even if those taxes do pay for basic services that people need, such as garbage pick-up, drinkable water, schools, and so on. Governments can often seem intrusive, taking your money and telling you what to do instead of the other way around.

Ever the optimists, we like to take a more positive spin on governments. A government is the servant of the people in a democracy. In the battle against global warming, governments are invaluable allies in stopping — and reversing — climate change. The Intergovernmental Panel on Climate Change (IPCC) reports that reducing greenhouse gases (GHGs) as much as the planet needs requires government leadership in the form of regulations and programs.

THE NONPOLITICAL POLITICAL ISSUE

Climate change has become a hot political issue. Depending on the political culture of your country, the issue is nonpartisan, pan-partisan, or more than a little partisan. In some countries — such as the United States, Canada, and Australia — the conservative versus liberal fight on climate change has never been more heated. The following provides a snapshot:

- **Australia:** In 2007, the defeat of John Howard's government was the first election globally in which climate change was a significant factor. The next government changed policy to be pro-climate action, only to have that climate action canceled when the right-wing party came back to power.

- **Canada:** Canada's Liberal government was defeated in 2006. The Conservatives under Stephen Harper canceled Kyoto. After nine years in office, the Conservatives lost to the Liberals who in 2015 proclaimed "Canada is back!" The new Trudeau Liberals in power signed on to the new Paris Agreement.

- **United States:** Much the same kind of whiplash policy occurred when Barack Obama and a (fairly weak) pro-climate action policy was replaced with Donald Trump, who pulled the United States out of the Paris Agreement. Then, more whiplash, when Joe Biden came into office and rejoined Paris, moving the United States back to more climate ambition.

In Europe, however, the fight against global warming isn't as divided along party lines. In the U.K., when the social democratic (Labour) government was in power, it passed a Climate Accountability Act to require the government to reduce emissions based on a carbon budget. This law was respected and continued to achieve carbon cutting goals when the conservative (Tory) government under Theresa May took over. The U.K. government continued following the same climate policy. The Climate Accountability Act continued slashing GHG emissions.

Perhaps the most influential conservative leadership on climate change has been from German Chancellor Angela Merkel. She was Minister of the Environment in the early 1990s when the mandate to negotiate the Kyoto Protocol was just beginning, and Merkel surprised some observers by maintaining the previous left-leaning government's course for GHG reductions when her party came to power. She also led the European Union in the same year that she helped push for major reductions in energy use. In 2007, she led G8 discussions, keeping a focus on climate change within the G8. As she left power in 2021, she continued to press the world to more ambitious climate action.

The global COVID pandemic has changed many peoples' opinions about what government can do. Many people looked to government — not private corporations — for emergency income support, public health measures, and pressure on pharmaceutical companies to develop vaccines and then supply adequate supplies of vaccine to protect populations around the world. Government action, while imperfect, was essential in saving lives.

In this chapter, we take a look at how governments can fight — and are fighting — climate change through initiating programs and regulating and taxing emissions. Sometimes, governments can help when they get out of the way. The entrenched interests of the status quo — also known as fossil fuels — have governments making promises out of both sides of their mouths, promising to solve the climate crisis, while also supporting fossil fuels. Market forces that push toward renewable energy are blocked when governments keep subsidizing fossil fuels. Climate crusader Bill McKibben calls this the "First Rule of Holes: If you are in one, stop digging!"

Knowing what solutions governments can enact empowers you as a voter; you can knowledgably select candidates whose climate change plan seems like it'll be the most effective. Beyond that, by letting candidates know that you're voting for action on climate change, you'll encourage all candidates to give this crucial issue a higher priority. We also share some success stories, so you can see what your votes can do!

If They Had a Million Dollars . . . Wait — They Do! — Funding Measures

When it comes to cutting GHG emissions, money is a powerful tool. Governments can create ways that people and companies benefit financially by making the right environmental choices. These incentives can take the form of tax credits and rebates, or governments can directly fund initiatives that plan to cut GHG emissions. Governments can also tax — as you probably know all too well. Taxes flow to government coffers for public services. But tax policies can also send a message — by what gets taxed and what gets subsidized. Governments can make reducing carbon-based fuels a smart financial decision by taxing carbon. These sections take a closer look at some of the different ways that governments can fund the fight against climate change.

Creating incentives

An excellent way for governments to encourage proper behavior (climate change–related or otherwise) is to offer *incentives* — a sort of reward that encourages people (or organizations) to act in a certain manner. You were probably offered incentives as a little kid to get good grades or tidy your room. (Governments are likely to offer more than chocolate sundaes to entice people to help fight global warming, however.)

TECHNICAL STUFF

Incentives can come in many forms: a direct cash return or rebate, a tax break, or a lower consumer price thanks to subsidies to the manufacturers. Not all subsidies and incentives are announced and heralded. Some are embedded deep in a tax system and allow corporations to write off costs and reduce the taxes they have to pay.

Incentives are popular with government officials who would rather reward good behavior than risk alienating the electorate through regulations and taxes. Observers such as the IPCC argue that incentives are most effective when governments use them in combination with regulations, which we talk about in the section "Laying Down the Law," later in this chapter. Otherwise, it's a case of too much carrot, not enough stick.

Here's a list of incentives that the IPCC recommends:

>> **Agriculture:** Most governments already subsidize agriculture; to help combat climate change, governments can subsidize farmers who take active steps to reduce their GHG emissions by such measures as improving land and soil management and using efficient farming methods. (We talk about how farmers can go green in Chapter 14.)

>> **Consumer goods:** Because low-emission products can sometimes cost a little more than inefficient wasteful ones, the government can encourage people to buy the more expensive product by footing some of the bill. A few examples include the following:

- **Canada:** The government offers grants of up to $5,000 so that owners can retrofit homes to make them more energy efficient.

- **Japan:** The Ministry of Economy, Trade, and Industry offers subsidies for homeowners buying automated systems such as thermostats and electricity regulators, which help reduce energy consumption.

- **United States:** The federal government offers rebates to anyone buying an EV (electric vehicle) or plug-in hybrid car.

>> **Forestry:** Subsidies to this industry could increase the amount of forest space, reduce or eliminate the logging of old-growth forests (which store far more carbon than newly planted forests), and encourage forestry companies to manage harvested forests in a sustainable fashion, keeping the ecosystem and soils healthy. Forests contain nearly 90 percent of all the carbon stored on land (known as *global terrestrial biomass*).

Forest management is carbon management. Recent research from the University of Hamburg confirms that 70 percent of the carbon a tree stores is stored in the last half of its life. Forests also store carbon in the soils. (Check out Chapter 2 to find out why trees and soils are important to the climate.)

>> **Waste management:** National or regional governments could offer financial incentives to encourage good waste management practices within their cities and communities. These incentives could include encouraging the diversion of waste from landfills through reduced packaging and recycling. For the remaining waste, initiatives could include incentives to capture and/or use methane-containing landfill gas, or funding local purchases of technology that breaks down waste more effectively than landfills or processes the methane gas emitted from landfills into clean electricity.

Planning for emissions trading

Emissions trading, sometimes referred to as *cap and trade,* is a complex and controversial approach to reducing GHGs. In this system, a government decides how much carbon a geographical region (such as a city) or a sector (such as all facilities that generate electricity from coal) is allowed to emit. The allowable level is called the *cap.* The trading comes in when an emitter produces a level of carbon less than its cap — it can sell the difference to another emitter that has gone over its limit. The promise of the revenue from selling their extra credits encourages companies to accelerate the adoption of new low-pollution technologies. Over time, the government lowers the cap, gradually reducing the allowable credits. The intended result is less and less pollution.

Here are some examples where emissions trading has had some success:

>> **Los Angeles:** Within North America, non-GHG emissions trading has had some historical success on other environmental issues such as air pollution. Los Angeles improved air quality through trading nitrogen oxide and *volatile organic compounds* (liquid chemicals that can immediately evaporate into gas at room temperature) credits, according to Western Climate Initiative (WCI).

>> **California:** California continues with its trading approach through WCI. The trading system uses emissions trading in a major carbon market involving Canadian provinces Quebec and Nova Scotia.

>> **Europe:** Europe has had an active GHG market since 2005. It was the first market of its kind to span different countries and numerous sectors. The EU carbon market is the model used around the world for building a market that trades carbon dioxide emissions and credits.

The following sections delve deeper into emissions trading and discuss the pros and cons and the debate that has been ongoing to find some consensus.

Focusing on the pros and cons of emissions trading

Emissions trading has its upside and its downside. On the one hand, it regulates emissions measurements and enables governments to hold companies accountable for the measured emissions. On the other hand, putting a price on carbon is a thorny matter. Do governments charge for only the carbon being directly emitted by burning coal? Or do they put a price on that *and* on the emissions produced by mining out the coal and shipping it?

WARNING

Emissions trading is often attacked as a license to pollute. The European Union Emissions Trading System, for example, has been in place for four years, yet countries such as Poland and Germany are still building coal mines because the added cost of the emissions trading is so marginal that it doesn't influence their decisions. Thus, emissions trading can actually end up being ineffective if the price is set too low, according to Inside Climate News. Market mechanisms can work, but only if the price is effective and governments monitor trading to prevent cheating and steadily reduce the cap over time.

Industries can engage in emissions trading without government involvement, too — that is, they can limit their net GHG emissions on a voluntary basis.

Meanwhile, energy from new solar installations is already cheaper than energy from existing coal-fired power plants and is pushing coal out of the market. Renewable energy is rapidly becoming less expensive than *any* fossil fuel system. This lower cost will lead to the replacement of most fossil fuel energy with cheaper, more efficient, and much more environmentally friendly renewables. Governments can assist or resist this revolution, but the transition will happen with or without them — those that choose to encourage the transition will benefit more and earlier than those that resist, but either way those nations will also make the shift. (We talk more about this inevitable massive disruption in Chapter 17.)

Looking deeper into the debate about emissions trading

When trading emissions within countries, a key sticking point is something called "additionality." Can the countries benefiting from the trades provide assurance that the emissions reductions wouldn't have happened anyway? In other words, are these carbon reductions *additional* to the status quo?

Debates about a global emissions trading system continued to be the most controversial part of the 2015 Paris Agreement with the attention on Article 6. At every annual meeting after COP21 in Paris, the attendees have tried to resolve Article 6. Many groups and countries argued, and for good reason, that a bad version of Article 6 — allowing fake trades that didn't really reduce emissions — was worse than no Article 6 agreement at all. Through COP22 in Marrakech, COP23 in Bonn, COP24 in Katovice, Poland, and COP25 in Madrid, agreement eluded the frustrated negotiators. Then in 2021, at COP26 in Glasgow, a delicate balance was achieved.

The solution lay in learning lessons from the system that had been adopted under Kyoto. In 1997 in Kyoto at COP3, countries decided to allow wealthier countries to provide funding for carbon-cutting measures in poorer nations, and then claim the emissions reductions as though they had been achieved in the wealthier country. This was called the Clean Development Mechanism (CDM).

The CDM accumulated a long list of dodgy trades — credits that appeared to be fraudulent. Wealthy countries were benefiting from credits that were based on faulty accounting for carbon emitted.

The Glasgow Rule Book for Article 6, referred to as Internationally Transferred Mitigation Outcomes (ITMOs), set out three new ways to trade. The trade in ITMOs involves two kinds of approaches:

>> Voluntary agreements between two countries that can prove they're actually reducing emissions

>> Trades encouraging private sector investment

The three new ways to trade focus on the following:

>> **Cooperative approaches:** Parties in bilateral arrangements recognize the transfer of emission reductions between them. This enables mitigation programs like emission trading systems in countries to link to each other. The parties can develop their own rules as long as they're consistent with the United Nations rules.

>> **Rules, modalities, and procedures:** The attendees adopted rules and procedures for the new UNFCCC mechanism, which credits emission reducing activities. Through these new rules not only nations, but also private sector companies can engage in global trading. This essentially creates a value in trading on a market basis — reductions in one place that a company can then sell to a private sector company in another country.

>> **Nonmarket cooperation:** Ideally this part of the agreement will allow countries, institutions, and other stakeholders to launch cooperative efforts to reduce emissions.

Putting Programs into Place

One way governments can fight climate change is through engaging and empowering their population. As we discuss in the following sections, by supporting research, educating people, and helping people adapt to the changes that global warming will bring, governments can put the tools for survival in the public's hands.

Research

Government funding for energy research skyrocketed in the mid-1970s, during the oil shock (when crude oil prices rose sharply), to encourage scientists and companies to develop alternative energy sources. But when the first crisis faded, so did interest. Government investment in energy research around the world is half of what it was in 1980.

Now that the world is facing a *climate shock,* oil prices are down from their previous peaks (but periodically come back up again in response to temporary shortages), and new global investment is moving to renewable energy. Still gaps do exist where more research is needed. The big oil companies, though, have been very profitable for many years and have enormous cash reserves. They like to keep that cash working where they're happiest — back in oil and gas, so they're investing lots of money in carbon capture and storage, with the intention of reducing the amount of GHGs produced for each barrel of oil (while, of course, they still want to increase overall emissions by increasing the numbers of barrels). The industry makes a great deal of noise about their investments in the "transition" away from fossil fuels, but in fact almost all their investments are in their existing businesses. More on this in Chapter 20.

The IPCC recommends that governments invest in two types of basic research to help reduce GHG emissions and develop renewable energy technologies:

>> **Private research** is done by companies. The downside is that the results are owned and sometimes even patented by the company, restricting access to the new technologies they develop. The upside is that results drive the research, and that research is often in line with the market and what consumers want — in other words, it gets the job done.

>> **Public research** is done by the government itself. The upside is that anyone can study and use the results, but public research often takes a long time because no market push drives the research.

The private sector and government researchers need funding to develop low-emission technologies at roughly the same rate that the climate is changing. *Low-emission technology* is essentially any kind of technology that reduces GHG emissions: It can be renewable energy technology, advanced battery technology, or more efficient appliances and lighting (see Chapter 4).

At this point in the development of new energy technologies, private sector investments dwarf those of governments. The business is now in a *virtuous cycle*, where the returns to new investments are so clear that new money is flooding in. Governments still have a valuable place in early-stage basic research, necessary to provide the underpinnings for further commercial work. And indirect government investment through subsidies and incentives can have a marked effect on later stage research and implementation of new technologies.

Education and awareness

Global warming is a complex issue — you could even write a book about it! Not surprisingly, people often have a lot of misconceptions about climate change and what they can do to help reduce their carbon emissions. Because not everyone's going to read *Climate Change For Dummies*, governments can play a big role in educating people so that those people can make informed choices.

GOOD NEWS

The United Kingdom's government has done an excellent job of educating its citizens about climate change. Here are a couple of the initiatives launched by the British government:

>> A national climate database (www.gov.uk/environment/climate-change-energy) — see Chapter 21 for information on this great resource

>> A large-scale, climate-awareness, art exhibition, featuring world-class photographers whose pictures portray the global impacts of climate change, as well as the various global tactics taken to reduce carbon emissions

One great way for governments to raise climate change awareness is to include it in school curriculums, creating a profound effect on society when these students grow up. Finding out about climate change at an early age could influence decisions these students make for the rest of their lives.

Of course, awareness campaigns don't necessarily reduce personal GHG emissions — in fact, ample evidence demonstrates that education and exhortation on their own produce very little effect. And even when personal habits are massively impacted, as in the COVID pandemic, with billions of people not driving and not flying, the impact on global GHG emissions was surprisingly modest — at 6.4 percent reductions as against emissions the previous year. Bottom-up individual actions help, but top-down decisions by governments are urgently needed.

Adaptation

The climate is already changing, in some areas more than others. It's going to keep changing over the next century. Even when humanity succeeds in slashing *future* emissions, much of the dangerous impacts is already certain because of *past* emissions. Governments can play a key role in dealing with the changes that civilization can no longer prevent by providing funding and resources to help people and businesses adapt.

Adaption starts at the local level

Climate change will bring profound changes to the globe as a whole — but the particular types and scales of these impacts will be profoundly affected by local conditions within your very own neighborhood. Because the changes people will face vary from place to place, local governments (city or regional) will be best equipped to address these problems — one-size-fits-all solutions won't work. National governments that signed and ratified the UN Framework Convention on Climate Change (see Chapter 11) agreed to undertake adaptation planning. Several countries, such as the Dominican Republic, Cuba, and others in the Caribbean, are also undertaking regional planning. But fundamentally, although climate change is happening globally, people need to react locally.

GOOD
NEWS

Miami-Dade County in Florida was one of the first communities to set out strong climate goals that local governments can do. In 2010, it created a plan to reduce emissions, and it also planned for how to adapt to changes it could not avoid, including sea level rise, salt-contaminated drinking water, and erosion. Miami-Dade led the way as a pilot project for Climate Resilient Communities, led by the International Council for Local Environmental Initiatives (ICLEI). (See the nearby sidebar for more about what local governments are doing.)

WHAT A LOCAL PARTNERSHIP IS DOING

The Local Governments for Sustainability (formerly the International Council for Local Environmental Initiatives — now known as ICLEI without reference to its original name) started as a partnership between 200 city and town governments from around the world in 1990. It has grown into a larger partnership of more than 700 governments that share a strong commitment to sustainable development, according to the International Institute for Sustainable Development. Leading cities include Austin, Texas; Iowa City, Iowa; Mississauga, Ontario; Makati, Philippines; and Guadalajara, Mexico. Their projects have a large range of objectives, including the following:

- **Cities for Climate Protection:** This campaign puts policies and practices in place to reduce the city or town's GHG emissions and improve the city or town's quality of life.

- **Sustainable Communities and Cities:** This program's goal is to help cities foster justice, security, resilience, viable economies, and healthy environments.

- **Water Campaign:** This program measures how much water a city or community is using, develops a plan that provides targets for efficient water use, works to meet those goals, and continually measures and improves those goals. Water use and infrastructure require a lot of energy. By conserving water, you can conserve energy and reduce GHG emissions.

- **Biodiversity Initiative:** This initiative increases the local government's role in conserving the diversity of plants, animals, and entire ecosystems within or around a city or community. In the same way that you try to keep your immune system healthy so that your body can deal with and recover from a fever, this biodiversity project keeps ecosystems healthy so they can better deal with climate change.

However, on the other hand, one tragic example of a failure to adapt is New Orleans. In light of global warming, the city council had just realized that they needed to take action to protect the city from extreme weather events. In the summer of 2005, the mayor spoke as a signatory to the Mayors' Climate Protection Agreement, which endorsed the Kyoto Protocol. (See the section, "Identifying Some Success Stories," later in this chapter, for more about this agreement.) He pointed out that New Orleans was one of the most vulnerable cities in the United States to severe weather events and would be even more vulnerable given the trends of global warming. The city council realized that they needed to repair the dikes and levees and beef up emergency planning, but Katrina hit before they could make the changes.

How regional governments can assist in adaption

Here are several examples of actions that regional governments can take to help their citizens adapt:

» **Diversifying crops:** Encouraging farmers to grow a wide range of crops helps governments protect both the food supply and farmers' incomes by ensuring that those farmers don't put all their eggs in one basket, so to speak. For example, if a pest that became rampant because of high temperatures causes the barley crop to fail, the farmers will still have the food and income from the corn and wheat crops that the pest didn't affect.

» **Expanding green space and tree cover in cities:** This measure absorbs runoff from large rainfall events and reduces local temperatures, lowers demand for air conditioning, and stores carbon dioxide within the growing trees and associated soil. (Flip to Chapter 2 for carbon cycle fundamentals.)

» **Improving floodwater management systems:** This involves taking extra measures to deal with more extreme and frequent flooding — such as making sure the street sewer system is working and maybe adding more street draining systems. Less asphalt and more green space in cities slows rapid runoff to storm sewers, reducing erosion and downstream flooding. Governments might also want to rethink and rebuild water treatment plants; the water systems were built for a different climate.

» **Improving or installing storm-warning systems:** This involves preparing for more intense and more frequent storms — for example, by establishing a way to notify everyone when a big storm is coming.

» **Improving water management and watershed planning:** Cities need to plan for more extreme droughts and more intense river runoff, and ensure that they'll have drinkable water available for their citizens.

» **Increasing disaster relief funds:** Governments need to put aside more money for cleaning up after events such as hurricanes and flooding.

>> **Securing levies against flooding:** Ensuring that the physical barriers protecting the city or town from flooding are strong and always being upgraded in advance of need enables cities and towns to continue to provide effective protection, even as the risk and intensity of flooding increases over the course of the 21st century.

>> **Setting building restrictions on coastlines:** Governments can prevent the development of buildings too close to the water's edge to avoid loss of property due to erosion, which will be accelerated due to rising sea levels that will come with climate change.

>> **Shifting tourism and recreation opportunities:** Tourism draws can be adapted to the changing climate and what the climate might be like in ten years. Some ski hills, for instance, may no longer be in operation because rising temperatures will rob them of reliable winter snow cover.

>> **Subsidizing farmers who may need to relocate:** Governments may need to support farmers who have to move unexpectedly because their crops are no longer growing in the changed climate.

The cost and effectiveness of planning for adaptation are uncertain, but governments can gain insight from other groups' experiences. Here are some examples:

>> Australia has been emphasizing water management.

>> Inuit communities in northern Canada are hunting differently to adjust to a changing climate and are maintaining their culture.

>> The Netherlands is building even more coastal barriers to protect against rising sea levels.

>> Ski resorts across the United States, Canada, Europe, and Australia are relying more on artificial snow to keep the slopes open.

REMEMBER

Adaptation isn't a response to just one particular storm or one melting glacier. It's a way of thinking, a kind of climate mindfulness that governments need to integrate into the plans they're already making. Australia already had a water management plan before climate change came along; climate change just added an extra angle — and urgency. Ideally, adaptation and mitigation (reducing GHG) goals should guide all policy. Many steps, such as urban tree planting and restoring mangrove forests, meet both goals.

HOW CITIES AROUND THE WORLD ARE SIGNING UP TO BE "RESILIENT"

Being prepared for what climate change has in store for cities around the world is about more than "adapting" to one massive extreme weather event after another. Yes, adaption matters. Changing the way highways are built to avoid washouts because culverts were too small for the massive increase in water volume, reinforcing seawalls and wharves to avoid being washed away or overwhelmed by storm surges, planning for cooling centers to protect people from the lethal effects of heatwaves, and thousands of other changes are part of what cities are doing to adapt.

Growing networks of cities are working together, sharing best practices, and working to both reduce local emissions and be better prepared for what it to come. Cities and their networks play a big role in what the United Nations calls actions by subnational governments. Leading international networks include the following:

- **C-40 Cities Climate Leadership Group:** 97 member cities, making up 25 percent of global economy

- **ICLEI – Local Governments for Sustainability:** 1,200 cities, towns, and counties

- **United Cities Global Governance:** 1,000 cities, 112 local governments

- **World Mayors Council on Climate Change:** 50 mayors

There are also substantial networks of cities in Europe, Asia, and the United States. One U.S. network started in 2017 when former President Donald Trump announced he would take the United States out of the Paris Agreement. "I was elected to represent Pittsburgh, not Paris," he said. The mayor of Pittsburgh, Bill Peduto, took offense. "Pittsburgh stands with the world and will follow Paris agreement," he said. Following Peduto's lead, mayors across the United States signed up to support Paris. More than 1,000 mayors from all 50 states signed on the U.S. Conference of Mayors.

Climate-committed mayors and cities are making big changes in how their citizens think about their future, including an attitudinal change in urban planning. Increasingly, people are looking to be resilient.

The Resilient Cities Network defines *urban resilience* as "the capacity of a city's systems, businesses, institutions, communities, and individuals to survive, adapt, and grow, no matter what chronic stresses and acute shocks they experience."

This network has member cities on every continent (except Antarctica) — from Accra, Ghana, and its need to confront water contamination and potential cholera threats, to Barcelona, Spain, with a growing reputation as one of the world's most livable cities, to Belgrade, Serbia, where on the banks of two rivers — the Danube and Sava — it faces economic and climate challenges, and to dozens of other cities everywhere. Resilience brings together strategies to survive and thrive.

Cleaning Up Transportation

Although governments don't control everything that has to do with transportation, they still have a fair amount of influence over it. And when it comes to cutting GHG emissions, transportation is one area in particular in which the industrialized nations can make dramatic changes — largely because they use some of the most inefficient forms of transportation on the face of the Earth!

REMEMBER

Different levels of government control different pieces of transportation. City governments, for example, control the public transit and city vehicles, which include public-service vehicles, such as park vehicles, snow plows, and garbage trucks. National governments can decide that a vibrant national rail system matters — moving people quickly and reducing GHGs. National governments can also control what kind of cars are sold in the region or in the country, where and whether highways or rail lines are built or upgraded, and how many lanes of traffic are dedicated to multiple-passenger cars on the highway.

In many modern cultures, especially in North America, people do think nothing about jumping in a car several times a day. Even for short errands, they automatically get behind the wheel. But there are choices. And those choices can be beneficial in more ways than one. Reducing GHG is healthy for the planet, lowering local air pollution is good for your lungs, and exercising — walking and biking — is good for fitness! These next sections explore some of those options.

Bringing back the bike

Many cities are recognizing that bicycles can provide an environmentally friendly alternative to cars on the road. Municipal initiatives to encourage cycling range from improving bike paths, adding bike lanes, and routing cars away from downtown areas. Many North Americans don't view the bike as a year-round transportation solution, but the Dutch and the Danes see things differently; a large percentage of city dwellers cycle contentedly throughout the cold and wet winters of northern Europe. Getting the right kind of bike and having dedicated bike lanes helps to extend the length of the bike-riding season in just about any city.

One of the world's most bike-friendly cities, thanks to its local government, is Amsterdam. There, you find separate bicycle lanes for each direction, with their own traffic lights that sync with those for cars, and spots abound where you can lock up your bike. City buses are equipped with bike racks, which passengers can use to stow their bikes when they board the bus. This sort of *intermodal transport* — switching between different ways to get around — is critical because it gives commuters convenient and flexible options, and makes it easy for people to not depend on their cars.

Investing in public transportation

Simply by providing adequate mass transit, municipal governments play a big role in reducing GHGs. Driving in stop-and-go city traffic produces the majority of car emissions. But governments can do more. Increasingly major cities are turning to purchases of electric buses — getting people out of their cars and into zero emissions vehicles. Edmonton, Alberta has purchased 40 electric buses — making it a national leader. In fact, the move to EV buses is picking up speed. The BC Transit Authority is committed to its entire fleet of buses (more than a thousand) being electric within 20 years. Montreal and Toronto have made similar pledges.

Many cities now have energy-efficient public transit buses. Your municipal government can run its public transportation fleets on alternative fuels or blends. In Halifax, Nova Scotia, whenever a bus goes by, you smell fish and chips coming from the tailpipe because its city buses run on recycled frying oil. (See Chapter 13 for more on alternative fuels.)

To get people to ride their energy-efficient public transportation, municipal governments can encourage riders in a number of ways. In London, youth under the age of 18 travel for free on major bus routes. Other regions create dedicated lanes on their roads to ensure a swift trip for public transit users. If it's faster to take the bus, why drive?

REMEMBER

Buses aren't the only form of public transportation. Trains are already one of the most efficient modes of transportation, and they have the potential to become 40 percent more efficient by using new technology, according to the IPCC.

Greening cars

Fossil fuel, which powers just about all vehicles, is the culprit for much of the world's GHG emissions. Increasingly, people are moving to electric vehicles (EVs). When we first wrote this book, there was a lot of hype around fuel-cell vehicles running on hydrogen. Fuel-cell vehicles may pull ahead again, but at this writing, EVs are winning the race for zero-emission vehicles on the show room floor. In

British Columbia, where Elizabeth and John live, one in ten new cars purchased are EVs — the highest rate in North America, according to Focus on Victoria. In Norway, thanks to government incentives, more than half of all new cars sold are electric. And China has more electric vehicles now than the rest of the world combined.

Even though EVs are taking off, at this writing only one in 250 cars around the world is electric. (Refer to Chapter 4 for more about fossil fuels.) Although governments can't force everyone into hybrid or electric cars, they can stipulate just how much fossil fuel goes into gas and how efficient cars must be at using that gas. Some governments have gone further, setting a deadline when internal combustion engine vehicles will no longer be sold. Japan, for example, has set 2030 as the year when all new car sales must be zero emission, per the International Council on Clean Transportation. Hertz, the car rental company, has recently ordered 100,000 electric vehicles. Other companies are expected to do the same.

The International Energy Agency (IEA) suggests that governments need to use subsidies and trade policies to increase the production of low-emissions fuels and technology. If your government adopts this approach, some of your taxes that are currently going toward oil and coal products would go to alternative fuel technology. We discuss how governments can fund research in the section "Putting programs into place," earlier in this chapter.

Dealing with personal vehicles

Many people have a love affair with their cars. Although we hate to break up a beautiful relationship between someone and their SUV, cars are big contributors to GHG emissions from transportation. Governments could definitely help nudge the auto industry and people toward making smarter transportation decisions. Meanwhile, *Car and Driver* reports that even big SUVs and truck models are increasingly available as EVs. Because of overwhelming demand, Ford has had to stop taking reservations for its new all-electric F150 pickup truck. Right now they're pretty pricey, but as demand increases, the prices will come down. Many Tesla drivers like the bumper sticker "Ask me if I miss gas stations." Zero emission vehicles are becoming cool.

But even charging stations for zero-emission vehicles take up urban space of value for parks and for outdoor pedestrian space. The COVID pandemic made many urban dwellers think differently about sprawling café tables in the middle of the street.

Cities could tackle the amount of space dedicated to parking. One significant improvement is to make parking spaces smaller. If your vehicle is too big, you pay for two parking spots, rather than one. Alternatively, cities could reduce the number of parking spaces they require in new developments — particularly if they're located close to transit services, increasing the value of parking spaces, and the convenience of not needing to find — and pay for — a parking spot. These and other measures could encourage people to drive smaller, more efficient cars. The city of Portland, Oregon, has gone one better. It virtually eliminated downtown parking spaces.

More cities could pass anti-idling bylaws, ticketing people caught leaving their engines running unnecessarily. The impacts of the pandemic have increased congestion in food deliveries to homes. All that commercial vehicle use will benefit the planet when they switch to zero emission vehicles.

Redefining Long-Term Investments

The government makes investments with its money just like you can. But the government doesn't put its funds into stocks and bonds. Rather, governments select industries or economic sectors that they feel benefit their citizens one way or another, perhaps in a perfectly straightforward fashion (for example, by providing jobs) or in a less immediate, more long-range way (by encouraging farmers to stay on the land, for instance).

REMEMBER

Most governments invest in (or *subsidize*) the energy sector. Typically, this investment means that they provide funds to oil companies, but governments could use subsidies to send the energy sector in very different directions. Governments could financially support companies that develop renewable sources of energy, rather than supporting those searching for fossil fuels such as oil. According to the IPCC, reducing fossil fuel subsidies is an effective way to wean the industrial world off its oil addiction. The IPCC warns that oil will remain civilization's primary fuel source as long as it remains subsidized by governments. The IPCC suggests eliminating these subsidies (along with establishing taxes or carbon charges on fossil fuels and providing subsidies for renewable energy producers) as measures toward shifting to renewable energy supplies. The call to end fossil fuel subsidies has been reinforced from some of the planet's leading economic institutions: the International Energy Agency, the International Monetary Fund, and the World Bank. Yet, globally, government money continues to flow to Big Oil.

According to the International Renewable Energy Association (IRENA), in 2020, a staggering $634 billion (U.S. dollars) was given by governments to various parts of the energy sector. Fully 70 percent of those subsidies went to fossil fuels. Only 20 percent went to renewable energy, with 6 percent going to biofuels and 3 percent to nuclear. The old phrase "follow the money" applies here. For governments that claim they're committed to cutting emissions, if you follow the money, it doesn't look like they mean what they say. Governments aren't doing what they could to speed up the transition away from the fossils; in fact, they're a big part of and cause of the delay. More on this in Chapter 20.

Considering the lack of progress

At the Pittsburgh G-20 Summit in 2009, world leaders promised to end fossil fuel subsidies. But, according to the International Monetary Fund (IMF), in 2020, taxpayers worldwide subsidized oil, coal, and gas to the tune of $5.9 trillion dollars.

A recent report from Oil Change International concluded that Canadian fossil fuel producers receive more public financial support than any others in the developed world, while renewable energy gets less government help in Canada than in any other G20 country. These big increases in fossil fuel subsidies, just as countries pledged to stop these perverse subsidies, show the extent to which vested interests will try to resist the changes they know are coming.

If governments want to reduce GHG emissions, they need to shift subsidies toward low-emission renewable energy. This shift would be more than an investment for the current citizens of the country; it would be an investment for future generations.

Focusing on the good news

Governments are already moving in this direction. The IPCC reports that coal subsidies have dropped in the last decade around the world. By shifting such subsidies to renewable energy sources, governments can help those sources develop — at no additional cost to the taxpayer. Many governments are investing in renewable energy, but it's often a question of one step forward, two steps back. For example, in 2010 in Australia, the Labor Party under Prime Minister Julia Gillard took power thanks to a coalition including Greens. A strong climate plan was adopted. Unfortunately, as soon as a Liberal government took over in 2013 under Prime Minister Tony Abbott, fossil fuels gained support once again.

In Germany, Chancellor Angela Merkel who served from 2005 to 2021 established a strong political commitment to fighting climate change. Merkel, herself a scientist before going into politics, understood climate was a significant issue. Germany was making huge strides in developing renewable energy, but then China and its industrial interests got access to German technology (by more or less stealing the intellectual property) and started manufacturing solar panels far more cheaply. This was good for the planet because the price of solar dropped dramatically, but it was hard on German investments. Now that a new coalition government has taken over, with Greens in the coalition, the German government with Chancellor Olaf Scholz is determined to improve its targets, but 2021 was unusually cold and demand for energy increased. What had been rapidly dropping levels of coal burning started increasing. What had been a growing surge in renewables fell back.

The co-leader of the German Greens, Robert Habeck, is heading a super ministry of Economy and Climate action. The government has declared it will make unprecedented investments in renewable energy, to completely phase out coal by 2030 and to cut reliance on natural gas.

Recognizing the countries leading the way

Measuring progress all around the world, the World Economic Forum reported that of 115 countries studied, only 13 have made consistent progress in reducing fossil fuels and increasing renewable energy. Those consistent climate champions are as follows:

- » Sweden
- » Norway
- » Denmark
- » Switzerland
- » Austria
- » Finland
- » United Kingdom
- » New Zealand
- » France
- » Iceland
- » Netherlands
- » Latvia
- » Uruguay

Here are a few of the highlights from the climate leaders:

- » **Sweden:** One of the first countries to put in place a carbon tax, Sweden continues to come out in the top spot. Sweden is one of the only countries that has high energy consumption and low carbon emissions. In 2020, 54 percent of its energy comes from renewable power sources. In 2021, Sweden produced the world's first steel made without fossil fuels, which could be a game changer.

- » **Finland:** Finland was the first country to apply a carbon tax; the national program called Energy Aid, specifically for companies and corporations, has been in place since 1999. It covered 25 to 40 percent of renewable energy and energy conservation projects. From that early start, Finland's renewable energy sector has taken off. Solar and wind power are ramping up, and there's excitement about the development of green hydrogen. By 2030, half of all energy used in Finland will be from renewable sources.

>> **Denmark:** Denmark's record is amazing. Over the last 20 years, this tiny country of 6 million people has cut its GHG emissions in half, while over the same time doubled its economy. It has set an even more ambitious goal for 2030 to cut emissions by 70 percent compared to 1990 levels. Meanwhile, its generous social programs are one of the reasons it's rated one of the happiest and healthiest countries on earth.

>> **Norway:** Norway and Denmark have teamed up to get maximum benefit from wind power in Denmark. When Denmark has too much wind-generated electricity, it sells it to Norway. Norway uses the excess wind energy for its grid, and what isn't needed right then for electricity is used to pump water up into storage areas. Using pumped storage like a giant battery, when it needs electricity, Norway opens the sluices, the water rushes down and turbines make clean, renewable hydroelectric power.

>> **United Kingdom:** In 2002, the government initiated a large-scale solar demonstration project, paying $63.6 million to install photovoltaic solar systems. These funds covered 55 percent of the costs for public projects and 40 percent for large companies. Building on the success of that project, the government committed $172 million dollars in 2006 toward energy-efficient buildings that generate their own power. By 2008, the UK became the first country in the world to put in place a Climate Accountability Act. Breaking the patterns of many countries, when the Labor government that passed the Act was defeated, the new Conservative government continued on the same path. Climate targets, measured in five-year increments, continue to be met.

We discuss specific success stories in different countries in the section, "Countries," later in this chapter.

Solar, wind, tidal, and geothermal energy (see Chapter 13) used to be futuristic ideas, but countries around the world are using them today. In 2005, these forms of energy produced 2.2 percent of the world's electricity — 18.2 percent when hydroelectricity (mostly old-fashioned large hydro) is added in. Today, according to the IEA, renewable sources light up 28 percent of the electricity grids of the world. In fact, during the COVID pandemic when every other kind of energy saw a huge drop in demand, only renewable energy saw demand rise.

The game changer has been the price of solar energy. Since this book first came out, the price of solar photovoltaics has plummeted. For the first time ever, for a new installation to generate electricity, the lifetime cost of a solar plant is way cheaper than coal. And, unlike oil, as people say, if you get a *solar spill* — it's — it's a nice sunny day!

The continuously decreasing cost and increasing efficiencies of solar, wind, and battery technologies (called SWB) is the trigger for worldwide disruption. Whatever governments choose to do, SWB sources will simply outcompete fossil fuels in most of the world. More on this in Chapter 17.

Laying Down the Law

Regulations and taxes: hardly the way to win new friends. But as controversial and unpopular as they can be, they're vital tools for governments that want to control GHG emissions. However, compared to other major environmental threats, governments have used regulations very lightly to protect the climate. When governments were dealing with leaded gas, they just used regulations. To ban asbestos, governments used regulations. To save the ozone layer (you guessed it!), governments used regulations.

Governments can regulate everything from the amount of energy your refrigerator can use, to how much insulation your house needs, to the amount of gas it takes to move your car a mile (or a kilometer). Taxes can be a powerful disincentive to individuals and companies; stop polluting, or you have to pay!

The next sections look at the way governments can act to combat climate change. Some ways involve using money as the carrot or as a stick. A lot of measures, often known as regulatory, can make a big difference.

Improving building regulations

The way houses and larger buildings are constructed is as much a product of policy as it is engineering. Building regulations apply to a building's structural elements, the materials used, their proper installation, and so on — but regulations can also control home energy use. By stipulating how the house is insulated and what sort of doors and windows to use, building regulations can determine whether your new home will be an energy guzzler or an energy miser over its lifetime. (We look at how buildings can be environmentally friendly in Chapter 14.)

GOOD
NEWS

When it comes to battling climate change, the IPCC reports that how people build, insulate, and heat their buildings has more potential to reduce GHG emissions than changes in energy, industry, or agriculture. With the right regulations in place, building-related emissions could be cut 30 percent by 2030 — without affecting the profitability of the construction industry.

THE HONOR SYSTEM: SELF-REGULATION

One very basic form of regulation is self-regulation, in which the affected parties police themselves. As you can probably guess, politicians prefer self-regulation to putting emission limits in place. Having someone volunteer to do something is much easier than having to tell them to do it.

The major benefit of this type of regulation is simplicity. First, industries promise to reduce their GHG emissions below a self-determined baseline before the reductions. Then, governments check to see whether they're being compliant. If they are, great. If not, well, better luck next time — so, self-regulation doesn't guarantee GHG emissions reductions. Compliance is voluntary, and industries can easily set their reduction goal at zero.

Although successes are possible, self-regulation usually doesn't work, and it hasn't been proven to work at all on a larger scale. China is attempting to make it work by partnering with its 1,000 biggest industries, which draw on a third of the country's energy. The industries will measure, report on, and reduce their energy use through conservation and energy efficiency. The government will support the industries by promoting energy-saving initiatives, monitoring the progress, and helping to develop the training and implementation of the projects. Although it's a step in the right direction, the program doesn't have any specific targets.

In addition to improving standard building features, governments could demand that new buildings include certain elements that would make those structures green (not to mention much cooler — or warmer, depending on the season). They could mandate that all new houses and other buildings include solar panels to help heat and cool them, or that contractors build *smart homes* — buildings equipped with automated systems that control heat and lights to conserve energy.

Regulating energy use

National and local governments hold great regulatory power over energy use. GHG emission reduction targets are set domestically. Under Paris they're referred to as *nationally determined contributions (NDCs)*. But the Paris Agreement itself sets the shared goal of holding to 1.5 degrees C or as far below 2 degrees as possible.

Here are a few ways that governments can encourage more responsible energy consumption through regulation:

>> **Clear labeling:** Governments can require producers to label all appliances and mark how much energy they use, and — most importantly — governments can decide what level of energy use qualifies as efficient. A producer would have to meet that requirement before selling their products as efficient.

>> **Measuring usage:** Measuring is the first step to setting a target for reduction. California has installed smart meters in homes, which allow electricity consumers to see on an hourly basis how they're doing in their conservation efforts. Studies have proven that when people can see how much energy they're using, they cut back on their consumption. Watching the costs add up helps people remember to avoid *peak demand* periods (when more consumers are pulling power at the same time) for some energy-draining activities.

>> **Setting quotas:** Governments can stipulate that major users of energy purchase a certain percentage of their power from renewable sources. Australia, for example, requires major buyers of electricity — such as industries — to buy a portion of their energy from wind farms and solar-power producers. If a user doesn't meet the quota, it must pay a fine.

>> **Setting targets:** Governments can encourage the development of renewable energy sources by setting targets for the entire country. Under the Paris Agreement, each nation sets its own specific target — called the NDC — within an overall goal to hold the global average temperature to as far below 2 degrees Celsius as possible, and preferably to no more than 1.5 degrees. Setting goals for renewable energy are frequently part of NDCs.

Taxing the polluters

"There is no such thing as a good tax," Winston Churchill once said. But he also praised certain taxes. The most significant government power is taxation. Government uses taxes not only to raise revenue for the state, but also to discourage certain forms of behavior — think of the taxes on cigarettes and alcohol. Governments increasingly apply a similar tax policy to GHG emissions. Carbon pricing is now used around the world.

Here we take a closer look at carbon pricing and identify some countries and states using carbon taxes.

The ins and outs of carbon pricing

The economic impact of such taxes has been minimal, even positive, as long as this fee is offset by reducing taxes on other things, such as income, jobs, and profit. This tax reassignment is called *tax shifting*. It's not about more taxes; it's about different taxes.

Most economists agree that this sort of tax, commonly called a *carbon tax*, is the most sensible and cost-effective approach to reducing GHGs. Carbon taxes are now endorsed by the World Bank, the IMF, and the IEA — not to mention *The Economist* magazine.

TECHNICAL STUFF

The "carbon" in carbon tax can refer either to carbon dioxide emissions alone or to *carbon dioxide equivalents* — meaning all GHG emissions measured as a factor of carbon dioxide. The carbon market in Europe, for example, considers only carbon dioxide. New Zealand uses a carbon market that includes carbon dioxide equivalents. Both systems work, but those that use carbon dioxide equivalents can potentially have a greater effect because they include all GHGs.

WARNING

The IPCC warns of a couple of problems associated with taxing GHG emissions: The taxes can't ensure actual emission reductions (polluters may be perfectly willing to pay to keep on polluting), and the taxes are tough to put in place from a political point of view because few things are as unwelcome to voters as taxes. On the other hand, carbon taxes can be implemented quickly, and are simple to administer, both of which make them attractive to governments gutsy enough to introduce them. Taxpayers are more open to carbon fee and dividend where every dollar of *fee* or tax collected goes right back to the citizen as a dividend. Canada is unveiling this approach.

Carbon taxes in action

The most competitive economies in the EU have carbon taxes. So far, carbon taxes have spread from Europe to the rest of the world.

Here are some other examples:

>> **Africa:** South Africa was the first African nation to adopt carbon taxes.

>> **North America:** Canada, at both provincial and federal levels, puts a price on carbon. So too does Mexico and parts of the United States.

   At the state level, California, Oregon, and Washington have carbon taxes, as do some cities. Hawaii adds a surcharge to every barrel of oil shipped into the island state.

>> **Northern Europe:** Sweden has maintained its tax on pollution, increasing the tax as the amount of pollution drops. As a result, Sweden now has the highest carbon tax on the planet — at $126 US per ton of carbon. GHG emissions have dropped and the economy has continued to thrive.

Norway has reported great success in using taxes to reduce both carbon dioxide emissions and HFC/PFC (the most powerful of the GHG) emissions.

>> **Western Europe:** London's levy on cars entering the downtown area has been a very successful measure to reduce GHGs, improve air quality, and relieve downtown congestion. The initial protests let up because Londoners agreed that the city was vastly more livable with fewer cars clogging the heart of the city.

Identifying Some Success Stories

Throughout this chapter, we talk about what governments can or should do to help reduce GHGs. Don't get us wrong, however — governments all over the world, at every level, are already doing leading-edge work, moving toward low-carbon technologies and ways of life. Here we take a closer look.

Cities and towns

You hear "Think globally, act locally" a lot these days — you need a big-picture perspective, but you have to make changes on a small scale, within your own city or town. Cities in industrialized countries can really reduce local GHG emissions, as many cities have already discovered.

Cities often act as a team to support each other in climate initiatives. In 2005, Seattle mayor Greg Nickels launched the Mayors' Climate Protection Agreement to reduce city emissions by an amount equal to what the United States negotiated at Kyoto. As Kyoto fades and countries strive to meet Paris, the Mayor's Climate Protection Center continues the work with a burgeoning membership of more than a thousand U.S. mayors.

You can get a taste for what some of these cities are doing to reduce GHG emissions by checking out Table 10-1.

TABLE 10-1   **Low-Carbon Cities**

City	2021 Population	Successes	Targets
Austin, Texas, United States	905,807	Cutting energy emissions 8 percent in five years; Improving energy efficiency by 7 percent. Saving $200 million through energy conservation.	Net zero by 2040, equitably.
Berlin, Germany	3,640,000	Cutting GHG emissions 41 percent (of 1990 levels) achieved ahead of schedule by 2020. Saving $2.7 million a year.	Cutting GHG emissions 70 percent (of 1990 levels) by 2030. Achieving carbon neutrality by 2045 at the latest.
Cape Town, South Africa	4.7 million	Supplying clean and reliable energy to low-income households; Initiating top-of-the-line community housing project as part of the former Clean Development Mechanism of the Kyoto Protocol.	Mitigate at least 80 percent of emissions (against 2016 levels) by 2050. Net zero targets for new buildings by 2030. Zero emission buses by 2025.
London, United Kingdom	9,000,000	Cut 20 percent below 1990 levels by 2020. Reducing transportation carbon dioxide emissions 19 percent in one year. Making $343 million in one year through fees.	Net zero by 2030.
Melbourne, Australia	5,078,000	Cutting GHG emissions from council operations 53 percent from 2013–2019. Purchased 100 percent renewable energy from Melbourne Renewable Energy Project.	Zero net emissions by 2040.
Toronto, Canada	6,254,571	GHG cut by 37 percent below 1990 levels by 2018.	Goal is for GHG cut by 65 percent against 1990 levels by 2030 and net zero by 2050.

States, provinces, and territories

Governments at the state, provincial, or territorial level have a very important role in fighting global warming. Regional governments (sometimes called *subnational governments*)) that have had success tackling their GHGGHG emissions have emphasized renewable energy standards, establishing programs to encourage energy efficiency and creating cap and trade programs for industry.

Like cities, some regional governments have found strength in partnerships. The U.S. state of California and the Brazilian state of Sao Paulo decided to work together to fight GHG emissions because, as their climate agreement notes, they're in very similar situations. At the Montreal COP in 2005, they committed through a

Memorandum of Understanding to shared climate action, and they are still working collaboratively making lots of progress. Both states

>> Are major contributors to their respective national economies and have the largest state populations in their countries (California at nearly 40 million and Sao Paulo at 44 million).

>> Have the highest energy use in their countries.

>> Have suffered from major air pollution, which makes them interested in energy-efficiency measures for their regions.

>> Are regarded as leaders in developing and implementing programs for climate action.

>> Express frustration with the slow pace of action by their national governments.

California and Sao Paulo's agreement led to significant climate action. They were among the first jurisdictions in the world to pass laws requiring action to cut emissions. Both are credited with pioneering in areas of reducing pollution from cars, energy systems, and land–use patterns. The two states are working together on a number of projects. For example, California is working with Sao Paolo to implement a project to clean the air, using the same framework that California did with its Federal Clean Air Act. At the same time, Sao Paolo is working with California's planners to help replicate Brazil's successful Bus Rapid Transit in California. A lot of the partnership is based on sharing information — whether about ethanol, substituting diesel with natural gas, conserving state forests, or generating electricity from biomass.

The California Global Warming Solutions Act of 2006 required that the California Air Resources Board (CARB) meet specific and measurable commitments with firm deadlines. The CARB has used innovative measures that have been spectacularly successful. Communities set their own goals for reduced car pollution and better land use decisions. The CARB works to support those regional plans with a mix of carrots and sticks. Thanks to the California carbon pricing scheme of cap and trade, the CARB has funding for carrots! And by harnessing market mechanisms, CARB has developed a range of winning formulas.

THE EMISSION TERMINATOR: CALIFORNIA

California beat its target of reducing emissions to bring them back to 1990 levels by 2020 and did so four years ahead of schedule. California set a new target for a 40 percent cut against 1990 levels by 2030. (California is the 12th-largest emitter of GHGs in the world). The target is ambitious, but achievable.

Countries

Many countries have shown that nations can reduce energy consumption while growing economically. The key is to improve efficiency, doing more with less — or doing it differently. In North America, regional governments are taking the big steps. On a national level, European countries are leading the industrialized world in going green:

>> **Denmark:** Denmark significantly increased its use of renewable energy, with 50 percent of its electricity coming from wind, solar, geothermal, and bioenergy as of 2021.

>> **Germany:** Germany blew by its Kyoto targets and achieved 35.7 percent GHG reductions (as against 1990 levels) by 2018. The country is on track for 55 percent cuts by 2030. It achieved a new record of renewable energy power generation in 2018, with 85 percent of electricity coming from renewable sources.

>> **Switzerland:** A world environmental leader for a long time, Switzerland recently hit a speed bump on climate action. In 2021, voters rejected the next stage of commitments to cut its emissions to 30 percent below 1990 levels by 2030.

However, progress had been strong through a number of initiatives:

- Passing a federal law to cut methane emissions from waste

- Legislating a cut in carbon dioxide emissions from energy to 10 percent below 1990 levels by 2010

- Implementing an energy program, *SwissEnergy,* that works with cities to set energy-efficiency standards for buildings, appliances, and transportation, among other things (see the nearby sidebar)

- Providing *EcoDrive courses* that show people how to drive efficiently (thus burning less gas and creating fewer emissions)

- Negotiating voluntary agreements for carbon dioxide reductions, which include tax exemptions

The Swiss government has also taken a sustainable development approach to transportation. Thanks to Swiss government programs, major freight trucks have cut their annual road mileage by 6 percent while the freight volumes have increased. The money saved from this mileage reduction (over $1 trillion per year) is put into railway infrastructure. And because of this boost in funding, train travel is predicted to grow at least 40 percent by the year 2030.

Switzerland also boasts numerous projects run by private organizations. The Climate Cent Fund, for instance, takes a one-cent levy on every liter (¼ gallon) of gas that's sold, and literally every penny from this levy goes into a fund for projects that directly reduce emissions in Switzerland and around the world.

SWISSENERGY: HOW IT WORKS

A program launched by Switzerland's federal government, SwissEnergy describes its goals as "promoting energy efficiency and the use of renewable energy" to cut human carbon dioxide emissions. The program's strength comes from the cooperative approach of four departments (transport, energy, environment, and communications), which work together with cities, industry, organizations, and businesses.

SwissEnergy has cut Switzerland's consumption of fossil fuels while at the same time amassing 150 million French francs for climate action. Altogether, this fund has driven cuts of 16 million tonnes of carbon overseas and 2 million tonnes domestically. The use of renewable energies has increased, too.

You can find more information about SwissEnergy at www.bfe.admin.ch. (The home page isn't in English, but you can access an English version by clicking the English link in the upper-right corner of the page.)

Chapter **11**

Beyond Borders: Progress on a Global Level

The terms used to describe the effects of greenhouse gases (GHG) on the world's atmosphere are changing. People used to talk about *global warming*, and some now use the term *global heating*. But global warming simply isn't a complete description. The term *climate change* is more scientifically accurate. Increasingly, people talk about the *climate crisis* or the *climate emergency*. All are correct. The problem isn't just local or national, it's global. Humanity is experiencing a global problem — and that problem requires a global solution.

The United Nations has an extremely important part to play in fighting climate change. It provides a forum for governments to work together and hammer out solutions to international problems. Global problem-solving is a long, slow process — and a thankless one much of the time. You may have read news articles about the countless international conferences on global warming and wondered what goes on at those meetings. Different nations bring competing agendas to the table; representatives from all nations must overcome language and cultural barriers; and national governments face pressures at home from business, organized labor, and opposition parties. The world's glaciers may be receding faster than international agreements can move forward.

And yet, despite all the impediments, the world's nations are making progress. Even better, sometimes they enjoy huge successes, such as the international agreement to stop the destruction of the ozone layer (which we discuss in the sidebar "International agreements work: The Montreal Protocol," in this chapter). The world today is a safer place because of global agreements.

REMEMBER

Global agreements hold countries accountable for certain actions and give nations a set of rules enforced through U.N. international law.

The world's governments have been struggling with climate change for more than 40 years. The process has been painfully slow, and those governments still have a lot to do. But, right from the start, every country (well, almost every country) agreed that no one nation can solve the problem of climate change alone. In this chapter we go beyond the headlines you may read about these international agreements, and we explore just why they're so imperative and what goes into making these agreements happen.

Understanding Why Global Agreements Are Important

Countries can do a lot to tackle global warming individually, as we discuss in Chapter 10. But the problem is far too great, and the solutions are far too complex, for countries to attempt to address climate change on their own. Each country is responsible for a portion of GHG emissions and has the ability to reduce global emissions anywhere from a fraction of a percentage up to 25 percent. But it's only with a collective effort that global emission can be reduced 50 to 80 percent. The world needs a global agreement to reduce greenhouse emissions and fight climate change because such an agreement can

>> Ensure that all emissions are covered and no one jurisdiction can create a carbon loophole. *Every* country needs to agree to limit its emissions. That's particularly true in the case of the biggest polluters. Everyone must be in.

>> Coordinate everyone's efforts. A coordinated effort ensures that everyone's working toward the same goal, rather than charging off in all directions.

>> Create an accepted target. Taking a big-picture approach allows nations to get a more accurate assessment of the actual impact of their emissions. Countries can then work together to determine targets for reduction that can make a difference. (Check out the section "Looking At the Paris Agreement," later in this chapter, to see how countries set their targets.)

>> Ensure a level playing field. In this age of a globalized economy, countries' government officials can feel terribly insecure, worried that businesses might leave them for a more accommodating nation if they impose emission regulations. If all countries commit to reducing emissions, those government officials don't have to worry as much. (We talk about regulating emissions in Chapter 10.)

>> Include the developing nations. Because developing nations have limited financial resources, they can't undertake initiatives for sustainable development or adaptation independently. A global agreement allows for wealthier nations to help these countries prepare for the effects of climate change and industrialize in a way that won't contribute more greenhouse gases to the atmosphere. (Flip to Chapter 12 for more about the impact global warming has on developing nations.)

>> Increase the transfer of technology and experience. When one country reduces emissions, it can share those best practices through international systems set up within the Paris Agreement.

The U.N. creates a global sense of understanding, and it provides a structure and venue for countries to work on issues together. It's the only arena for global agreements because it was designed for that very purpose. And it gets results. (See the sidebar "International agreements work: The Montreal Protocol" for an example of a successful global effort to reduce emissions.)

Examining the U.N. Framework Convention on Climate Change

The year 1992 saw the biggest gathering of heads of government ever held — the Earth Summit in Rio de Janeiro, which drew together 170 countries and close to 20,000 people. Leaders rarely seen on the same stage, such as then-U.S. President George H. W. Bush and Cuba's then-president Fidel Castro, made first-of-their-kind commitments that continue to this day, on the topics of sustainable development, biological diversity, and — of course — climate change.

The summit's aim was to address the twin issues of environmental threats and poverty. Work began in 1990 to negotiate measures to which all nations could agree. The measures to fight poverty never came to pass. Two large environmental treaties did make it to becoming legally binding on most countries on earth. One dealt with protecting nature (the Convention on Biological Diversity). The other tackled climate change. The world leaders created the United Nations Framework Convention on Climate Change (UNFCCC). (In this case, a convention means a legally binding agreement — a statement of principles and objectives without specific numbers or measurable targets.) The Convention's main goal is to stop the build-up of GHGs in the atmosphere before the level of those gases becomes dangerous. (The Convention actually includes the word "dangerous" to describe the level of greenhouse gases that must be avoided.)

Of course, *dangerous* is a relative term. If you were in Europe during the heatwave of 2003 that killed 30,000 people, or in California fleeing the wildfires almost every summer since 2004, or in British Columbia escaping the devastating fires and floods in 2021, you might decide that climate change has already made the world pretty dangerous. In order to make the term *dangerous* a little less relative, the Convention *parties* (the participating countries) rely on the Intergovernmental Panel on Climate Change (IPCC). (We talk more about the IPCC in the section "Identifying the World's Authority on Global Warming: The IPCC," later in this chapter.)

Knowing how your local government passes by-laws is complicated enough. The distant, somewhat murky world of international law is way beyond what most citizens of the world experience. You might see it in movies — the United Nations towers on the banks of the East River in Manhattan, but what does it do?

The next sections set out the way the Climate Convention and since 1992 its subsequent more specific treaties have been negotiated and what they intend to accomplish.

Recognizing what the Convention does

The UNFCCC lays out the groundwork for action on climate change by

>> Committing all parties to a shared commitment to action (*committing to a commitment* is sort of like giving your sweetheart a promise ring pledging that you'll get engaged — you're promising to make a promise).

>> Acknowledging that climate change is occurring and that human activities, such as burning fossil fuels and changing land use (like deforestation), are the major sources of this change.

>> Adopting the precautionary principle as a basis for action. The *precautionary principle* calls for preventive action under conditions of uncertainty — no party can use a lack of scientific certainty as an excuse for inaction. The precautionary principle also puts the burden of proof on the proponents of an activity — it isn't the responsibility of society to demonstrate that something might be harmful, it's up to the proponent to show that it won't be. As time has passed since the original UNFCCC, this principle has failed. Countries have failed to act even as climatic events have become more dangerous, the science has become more certain, and industry has continued to cause harm.

TALKING THE GLOBAL AGREEMENT TALK

We talk a lot about conventions and protocols. And you've probably read about other conventions, too — the Geneva Convention, for example, on the rights and treatment of prisoners of war. In diplomatic terms, these words have very specific meanings — quite different from how people normally use them. In this context, a convention basically involves agreeing to a principle — in this case, fighting climate change — and setting objectives, but with no timelines or specific targets.

As soon as all the countries' government officials agree to the language of a convention, they normally sign it right away. Then, they take it back to their countries, where it must be ratified, or domestically approved. Ratification processes vary. In the United States, treaties must be subjected to a Senate vote where three quarters of the Senate must approve before the U.S. is bound by international law. In Canada and other parliamentary democracies, a treaty can be ratified by a simple Order in Council within Cabinet.

After a country ratifies a convention, that country becomes a party to that convention.

All the countries on earth have ratified the UNFCCC. All 200 nations are parties.

Conventions generally agree on a ratification formula — the number of countries that need to sign on before the convention is enacted (or, in U.N. language, enters into force) and becomes legally binding.

Later, parties can agree on a protocol or agreement, an agreement that's stricter and more detailed than the original convention. Like a convention, it's legally binding, but it contains actual deadlines and targets.

Establishing a game plan

The UNFCCC doesn't explicitly spell out how the parties involved should tackle climate change (that came later — we get into the detailed protocols or agreements that came from the UNFCCC in the section "Looking at the Paris Agreement," later in this chapter). Instead, the Convention committed the parties to "aim toward" stabilizing the level of GHGs in the atmosphere. It set out two areas in which the parties need to act:

>> **Adapting to climate change that can't be avoided:** The Convention acknowledges that, regardless of how much GHGs levels drop, due to the increased GHG concentrations from human activity, the world can't avoid some effects from climate change. All countries are going to have to adapt to

a changing climate regime. For example, some countries may need to plant drought-resistant crops; others may need to build higher levees and dikes in low-lying areas, and not rebuild on flood plains. (We talk about adaptation in Chapter 10.)

>> **Reducing GHGs:** The Convention calls this process *mitigating,* which means cutting emissions.

Dividing up the parties

The UNFCCC recognizes that industrialized countries have created most human-caused GHG emissions, and, therefore, it states that those nations should take the lead to battle climate change. This remains true even as developing countries like China and India have really increased polluting. Since pledging to cut emissions in 1992, even rich countries like the United States, Australia, and Canada have increased GHG emissions.

Countries vary in terms of how much they add to the problem and how able they're to actually help fix it. The UNFCCC breaks nations into three groups based on this variety, and it has different expectations for each group. In convention-speak, this division is called "common but differentiated responsibilities and respective capabilities." The following list describes how the groups are distinguished from one another:

>> **Annex 1 countries:** This group includes all industrialized (developed) countries such as Australia, Canada, the United Kingdom, and the United States. Annex 1 includes the following subgroups:

   ● **Economies in transition:** This group is made up of countries that are transitioning to a market economy — primarily former Soviet countries such as Hungary, Belarus, and Poland.

   ● **Annex 2 countries:** The Annex 2 countries generally have the strongest economies — this grouping includes all Annex 1 countries, *except for* economies in transition. Parties of the UNFCCC expect these countries to contribute money, technology, and other resources to Non-Annex 1 countries.

>> **Non-Annex 1 countries:** This group basically includes all industrializing (developing) countries, such as Brazil, China, and India. These poorer countries will have a much harder time adapting to the impacts of climate change than the wealthy industrialized countries in Annex 1 and Annex 2. They don't have the money for new technologies or programs, and they often have far more immediately pressing issues to deal with, such as war, famine, HIV/AIDS, or inadequate clean water. (We look into the challenges facing developing countries in Chapter 12.)

Historically, the greater a country's gross domestic product (GDP), the greater its volume of greenhouse gas emissions.

Looking At the Paris Agreement

Starting in 1995, after the UNFCCC was signed and ratified by enough countries to enter into force (EIF), the parties met annually to develop more detailed plans. Every year's meeting was a kind of climate congress. All the nations (parties) meet in a large gathering called the *Conference of the Parties* (COP). U.N. negotiations are always tricky because they require *agreement by consensus* — every party needs to be on board with the decision. Negotiations can go into the wee hours of the morning, often without a break. Bleary-eyed negotiators stay glued to their microphones while the translators share the discussions in six official languages.

The next sections review, in a very cursory fashion, over 30 years of climate negotiations. In 1992, Elizabeth was at that Earth Summit with her baby daughter who hadn't yet turned a year old. Looking back to Rio, she often tells people how that baby is now 30, working as a school teacher and on a PhD. She has, to put it mildly, made tons of progress! It's hard to see that much progress in the work of climate negotiators. Step by step, national governments have created a web of rules, to be replaced with newer rules. One thing that has always been lacking is a way to enforce the rules. And so the level of GHG keeps increasing. Nevertheless, the Earth would much worse off if governments weren't talking to each other and not committed to moving to protect a healthy planet.

Setting targets

National governments within the United Nations began discussing lowering GHG emissions back in 1990 at the first meeting to set up the UNFCCC. And until 2009 and the Copenhagen talks the conventions and protocols under the UNFCCC used 1990 emission levels as a shared baseline for setting reduction targets. That global norm was disrupted, first by Canada in 2006 and then by the United States at Copenhagen. As a result, the approach in the Paris Agreement was to remove targets and timelines from the agreement altogether. Instead of targets set out in the text of the agreement, each country picks its own target and timeline — referred to as a nationally determined contribution (NDC). These NDCs are filed with the secretariat of the UNFCCC. NDCs can be removed and replaced at any time, but only to ratchet up. In other words, targets and plans can change, but only to do *more* to fight climate change and help prepare for climate impacts that can no longer be avoided through adaptation plans.

REACHING AGREED-UPON MILESTONES

Through the last 26 COPs, 1995 to 2021, a series of protocols and agreements have been negotiated with many more detailed sub-agreements. Here are three notable milestones:

- **COP3:** At COP3 the Kyoto Protocol was negotiated, which set out specific reductions for Annex 1 (industrialized) countries against specific percentages and against the same timeline-reductions against 1990 levels achieved between 2008 to 2012. The Kyoto Protocol was undermined because the United States refused to ratify it. Canada ratified it, but after a change of government, legally withdrew. Despite those two countries, most of the countries who ratified it met their commitments. Unfortunately, global emissions still increased overall.

- **COP15:** Although a replacement for the Kyoto Protocol was attempted in Copenhagen in 2009, it ultimately fell short of support but continued forward into two further COPS, concluding an agreement at COP17 in Durban. Copenhagen was described as "politically binding" — a long way from the "legally binding" language in Kyoto. COP15 changed the base year to reductions against 2005 levels; however, specific reduction targets weren't included in the agreement, as in Kyoto. Instead, COP15 expressed goals to avoid going above 2 degrees C and preferably to hold to 1.5 degrees. Many countries remained outside the agreement.

- **COP21:** The most comprehensive agreement was negotiated — the Paris Agreement.

Each one of these was negotiated in good faith, but, unlike the Montreal Protocol on the Ozone Layer, climate agreements didn't have sanctions or penalties to enforce its provisions.

REMEMBER

Rather than using language of percentages and deadlines like in the Kyoto Protocol, the goal of the Paris Agreement is expressed in terms of global average temperature increase.

Under Paris, every country on Earth has committed to reduce emissions and enhance carbon sequestration in order to hold global average temperature increase as far below 2 degrees C (3.6 degrees F) as possible and to pursue efforts to limit temperature increase to 1.5 degrees C (2.7 degrees F).

Copenhagen and Paris – failures and breakthroughs

The focus on global average temperature increase took hold in the lead-up to the Copenhagen negotiations in 2009. Looking back at the 1992 UNFCCC, the goal was to avoid the build-up of human-caused GHGs (*anthropogenic* sources of GHG) to levels that could be considered dangerous.

The goal wasn't expressed as percentage reductions or global average temperature. As one negotiated plan after another failed to deliver results, scientists began to focus on what level of global average temperature increase would be dangerous. Many developing countries looked at the science and realized that they wouldn't survive on a planet with 2 degrees C (3.6 degrees F).

In 2009 at the Copenhagen negotiations, all the countries of the African continent and all those representing low-lying island states staged a dramatic walk-out, chanting "One point five to stay alive!"

The Copenhagen Accord — a thin document with weak buy-in — set as goals staying below 2 degrees C (3.6 degrees F) and trying to stay at no more than 1.5 degrees C (2.7 degrees F).

When the Paris Agreement was negotiated at COP21, the big question in those talks was whether 1.5 degrees would be included as a more ambitious goal than staying well below 2 degrees. In the end, 1.5 did stay alive and make it into the agreement.

A vast difference from a small change

But then, the negotiators and governments realized the difference between 1.5 degrees and 2 degrees (3.6 degrees F and 2.7 degrees F) wasn't really known. The IPCC was mandated to do a major review to answer that question and to deliver it before COP24. In October 2018, just before COP24 in Poland, the IPCC delivered its Special Report on 1.5 degrees. The news was sobering. The difference between 1.5 and 2 degrees could be measured in millions of deaths, millions of displaced people and refugees, loss of ecosystems, and extinction of species. Fortunately, the IPCC found that it was still possible to hold global average temperature increase to no more than 1.5 degrees (2.7 degrees F). To do so, carbon dioxide emissions globally must be cut 45 percent below 2010 levels by 2030.

That major IPCC report on 1.5 degrees led to a significant shift as more governments realized how much more dangerous 2 degrees was than 1.5 degrees. The concerns have mounted: Both the IPCC and the UN Secretariat for the UNFCCC take stock of the promises by governments. They assume all promises will be kept — not something that has occurred so far. Then they calculate whether the promises made will result in holding to 1.5 degrees. Unfortunately, commitments to reduce emissions on the trajectory required by the IPCC are (at this writing) still inadequate. When COP26 in Glasgow opened, the calculation of whether the total impact of promises would get the Earth to the essential goal (45 percent below 2010 levels by 2030) led to the dismal news.

WARNING

Just before Glasgow, keeping all promises, would lead to 16 percent *more emissions* than in 2010. After all the new promises from Glasgow were tallied up, world emissions would be 13.7 percent *above* 2010. An improvement, but still very much a path to failure.

Figure 11-1 shows the commitments made by some countries, and their actual performance to date. The Nationally Determined Contributions (NDC) column lists the amounts their governments have pledged to reduce emissions by 2030. The Measured Performance column lists the levels of GHG emissions (in thousands of metric tonnes) in 1990, 2005, and 2019. As you can see, European countries are progressing rapidly, and the United States and Japan are way behind but making progress, and Canada is the laggard of the bunch.

Nationally Determined Contributions — Emissions Reduction				Measured Performance (MT)				
Country	Base Year	Target Year	Reduction	1990	2005	2019	Change 1990–2019	Change 2005–2019
Canada	2005	2030	40–45%	602	735	730	21%	−1%
France	1990	2030	at least 55%	524	556	443	−15%	−20%
Germany	1990	2030	at least 55%	1249	993	809	−35%	−19%
Sweden	1990	2030	at least 55%	71	69	51	−28%	−26%
Japan	2023	2030	46%	1269	1370	1209	−5%	−12%
Norway	1990	2030	55%	51	55	50	−2%	−9%
United Kingdom	1990	2030	at least 68%	795	690	453	−43%	−34%
United States	2005	2030	50-52%	6643	7423	6558	−1%	−12%

Source: Organization for Economic Cooperation and Development https://stats.oecd.org/Index.aspx?DataSetCode=AIR_GHG

Adding flexibility

Before agreeing to the Kyoto Protocol and the Paris Agreement, some industrialized countries insisted on a number of compromises called *flexibility mechanisms* (or loopholes, according to many environmentalists). These mechanisms involve carbon credits. If a country lowers its emissions more than its target, it receives a carbon credit for the extra reduction. (It's similar to what happens if you pay more than you owe on your credit card bill; you get a credit that goes toward your next bill.) The country can decide to either apply this credit to the next commitment period (when it sets its second set of targets) or sell the credits to another country that's having trouble meeting its targets now.

INTERNATIONAL AGREEMENTS CAN WORK: THE MONTREAL PROTOCOL

In 1987, the United Nations Convention on Ozone met in Montreal, Canada, to negotiate a protocol to reduce the release of chemicals that were depleting the ozone layer — the layer of upper stratosphere that protects the Earth from the sun's ultraviolet rays. Many industries used these ozone-depleting chemicals as refrigerants and the propellant in aerosols.

The Montreal Protocol, a globally ratified agreement within the UN, acknowledged that industrialized countries had created most of the problem, that they had the best technology to solve the problem, and that poorer countries still needed access to chemicals so that they could economically develop.

The agreement required industrialized countries to cut production and use of ozone depleters by 50 percent, and less-developed countries could increase their use by 10 percent. Ultimately, 191 countries — almost every country in the world — agreed to get rid of ozone depleters altogether within a specified time frame. But the industrialized countries had to take the first step. The Kyoto Protocol takes a very similar, if not identical, approach.

The UN recently reported that ozone-depleting chemicals have been drastically reduced thanks to the Montreal Protocol. Recent studies show that the hole in the ozone layer has stopped growing and the layer is on its way to recovery. At the rate the hole is shrinking, the layer should be healed by 2050.

Because of the flexibility mechanisms, countries can actually make money if they lower their emissions more than they said they would. Of course, the downside of this trading is that some countries would prefer to buy credits, instead of actually reducing their own emissions. (Check out Chapter 10 for more details on carbon trading.)

Ratifying climate pacts

Each agreement has its own formula for ratification. For the Kyoto Protocol to become officially active, 55 countries had to ratify it, and those countries had to be responsible for at least 55 percent of the world's GHG emissions in 1990. Although negotiated in 1997, the Kyoto Protocol didn't become legally binding until February 2005.

When the Paris Agreement was being negotiated, the memories of the long and tortuous path of ratifying Kyoto were fresh in diplomats' minds. The 2015 Paris Agreement was completed with a parallel agreement to cover actions required before 2020 — just in case the agreement remained nonbinding for years. In the end, the Paris Agreement had so many countries ratifying that it took effect in less than a year.

International negotiations between 200 governments take a long time, but humanity doesn't have another way to solve complex global problems.

Introducing the World's Authority on Global Warming: The IPCC

Scientists around the world agree that global warming is happening due to human activity, but they don't all agree what its local and specific consequences will be. Thousands of studies are published every year that consider every conceivable aspect of climate change. To keep track of the top research, the United Nations formed the IPCC in 1988, which is a group of scientists appointed by their governments that assesses current research and compiles it in reports that it issues every five years.

In November 2007, the Nobel Peace Prize was jointly awarded to the IPCC and Al Gore for their work on climate change.

Some people, even in the media, report on the IPCC as if it were an advocacy group. Nothing could be further from the truth. In many ways, the IPCC is just as bureaucratic, requiring just as much diplomacy as the United Nations itself. It's also scientific. Top scientists are appointed by their governments. Because of that reality, that the word "intergovernmental" has meaning, it's also political. Negotiating text on when the world's scientific community agrees a threat is real at "very high confidence" is resisted by oil and gas–producing countries — and the fossil fuel industry. The next sections get into the work of the IPCC — and how that work gets done.

Getting to know the IPCC

The IPCC is made up of 2,500 scientific expert reviewers from 130 countries. These reviewers (all volunteers) come from a wide range of scientific backgrounds and include 450 main authors and 800 assistant authors who work together to create the IPCC's assessment reports. The governments of countries who are member parties of the UNFCCC can select scientists to be on the IPCC, often based on nominations from organizations or individuals. From these selections, the IPCC Bureau chooses their reviewers, picking candidates primarily based on scientific qualifications. However, other factors also come into play — the Bureau attempts to ensure that different regions, genders, ages, and scientific disciplines are all represented.

The IPCC's assessment reports offer a big-picture look at the science about, causes of, impacts from, and solutions to climate change. The IPCC's scientists look at all the relevant *peer-reviewed* science in the world (meaning any scientific article reporting on recent research about climate change that a group of scientists, other than the writers themselves, approve) and integrate this material into their report. The IPCC authors decide what to include by consensus, ensuring a broad and conservative agreement about what gets in. When disagreements occur, the IPCC notes them within the report.

Reading the reports

IPCC reports inform global talks to address the threat of climate change. The IPCC published the first assessment report in 1990, which played a key role in encouraging the attendees of the Earth Summit in Rio de Janeiro to create the UNFCCC in 1992. (The section, "Identifying the United Nations Framework Convention on Climate Change," earlier in this chapter, can tell you more about this Convention.)

The second report, completed in 1995, helped spur the Kyoto Protocol of 1997. After the third report was issued in 2001, those awaiting a scientific authoritative statement couldn't deny the man-made causes of climate change.

These IPCC periodic reports gained in certainty and scientific consensus of the links between human activity and the climate crisis. This edition of *Climate Change For Dummies* is relying on the most recent IPCC report. The Sixth Assessment Report of working group 1 (released in August 2021) was dubbed "Code Red for Humanity" by the U.N. Secretary General. It confirmed that emissions from burning fossil fuels and loss of forests was unquestionably causing dangerous climate conditions in every part of the world.

The IPCC's reports are heavy reading (literally — each report has about 3,000 pages and weighs approximately the same as a newborn baby). With each report, the connection between climate change and human activity has become clearer — and the warnings about what will happen if people don't act have become increasingly urgent.

The UNFCCC relies on the IPCC reports in their decision making. The reports recommend GHG reduction targets, regional adaptation strategies to climate change, and technological opportunities that can help reduce climate change. The reports' top-end advice is trusted by UNFCCC parties.

To find the main reports, visit www.ipcc.ch/ipccreports. On the website, you can select whichever type of report you are looking for. Most every report written by the IPCC has a "Summary for Policy Makers" or "SPM" version of the report that is free for download. This version, intended for policy people, not scientists, is a lot more readable than the main reports.

Chapter **12**

Developing in the Face of Climate Change

The political climate surrounding global warming is incredibly unfair. Although the major contributors to global warming have historically been the richest, most industrialized nations, now that those nations are waking up to the dangers of global warming, they're trying to hold developing nations to environmental standards that they themselves didn't face. Worse still, these developing nations face the same environmental challenges as other countries, but without the financial resources to prepare for them.

This chapter investigates the unique challenges that these countries face while they seek to develop their economies in the face of global warming. We look at some positive steps that China, Brazil, and India (three of the world's largest and most populous developing countries) are taking. Finally, we look at what initiatives developing countries can take to reduce their carbon emissions and adapt to a warmer world and how industrialized nations can help pitch in.

Identifying Challenges Faced by Developed and Developing Nations

The countries of the world are roughly divided into two categories:

>> **Developed:** These countries, which are also known as *industrialized countries,* have a strong industrial base and a relatively high income per capita. The generally accepted grouping of industrialized countries includes Australia, Canada, Japan, New Zealand, the United States, and countries within Europe. Many people would also consider countries such as Russia and Israel to be developed.

>> **Developing:** Also referred to as *low-* and *lower-income countries,* these countries generally have a low per-capita income and little industry. In developing countries, life expectancy is lower than in industrialized countries, and an increasingly urbanized population is growing more rapidly than the population in industrialized nations. All developing nations are moving toward industrialization (that's why they're described as developing), but some are closer than others. Most countries in the world are considered developing. Increasingly, the fast-paced growth of the economies of China and India make some people question if they really are developing.

Figure 12-1 highlights 46 of the world's least developed countries. They're broken down as such:

>> Africa: 33 nations

>> Asia: 9 nations

>> Caribbean: 1 nation

>> Pacific: 3 nations

The boundaries and names shown and the designations used on this map don't imply official endorsement or acceptance by the United Nations.

TECHNICAL STUFF

Some development academics argue that no country can ever truly be fully *developed* because progress has no real end. Others say that industrialization and a stable and strong economy signal that a country is developed. Another group thinks that development is defined by what educational and health services are available to the general population.

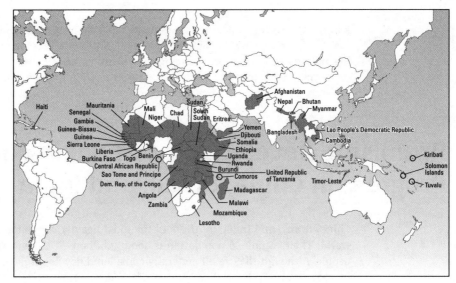

FIGURE 12-1:
The 46 least developed nations in the world, according to the UN.

Source: UNCTAD

Developing countries face significant challenges while they try to build and improve their countries' economies. The route to wealth that all industrialized countries followed (such as developing big industry by using fossil fuels like coal and oil) leads to climate change. As the wealth and industry in these huge developing countries grow, so too does their energy consumption. If these countries take their energy from the traditional sources that power industrialized countries, carbon emissions will skyrocket.

REMEMBER

Consequently, the development of these countries is under great scrutiny. Historically, some industrialized countries, such as the United States and Canada, have said that they wouldn't commit to reducing their impact on climate change unless major developing countries, such as China, Brazil, and India, also commit. But the "Do as I say, not as I do" position from major industrialized nations hasn't impressed developing countries.

Over the course of the negotiating process, from Kyoto through Copenhagen and finally Paris, all countries, developing and developed, have taken on targets and commitments. The fundamental principle, running through the successful Montreal Protocol on Ozone to the climate agreements is of "common, but differentiated responsibilities." This small phrase means that the community of nations agrees, "the rich go first, but we are all in it together."

The generally accepted idea, which is the core of the ozone and climate agreements (which we discuss in Chapter 11), maintains that industrialized countries caused the problem and have more resources to tackle it, so they should take the first steps in fixing it. After they get a good start, then developing countries can join in. The technological innovations of industrialized countries can help make the transition to a low-carbon development path more feasible, and the developing nations can focus on reducing emissions when they're financially able to implement new technologies, including receiving financial and technological assistance from industrialized countries.

Promising Developments: Looking at Progress in China, Brazil, and India

China, Brazil, and India are three of the most heavily populated countries in the world: Their combined population is about 3 billion — 42 percent of the world's people. Although they're all generally considered developing, each one is quickly moving toward industrialized status. Their development has led them down the same path that all other industrialized countries have traveled over the past 50 years — and industrialization, combined with growing populations, has led to massive increases in the amount of greenhouse gases (GHGs) they produce. China, Brazil, and India made up 34 percent of the world's carbon emissions in 2021. The rapid pace of industrialization in developing countries gives plenty of cause for concern.

REMEMBER

However, the climate crisis is the result of emissions over decades. Even though China's emissions have now surpassed those of the United States on an annual basis, the cumulative and historic emissions still have the traditional industrialized countries as having created the lion's share of the problem. And "historic" emissions aren't history. They continue to exert a warming influence on the Earth. Think of the atmosphere's capacity to hold carbon like a bathtub — the developed world's industrialization put most of the water in the tub there years ago, so the new emissions from China, India, and other developing countries is causing the overflow, when in fact they're still a small proportion of all that water put there.

In 2004, the United States, the United Kingdom, Canada, and Australia made up about 24.5 percent of the world's carbon dioxide emissions. However, this group of industrialized countries comprises just 6 percent of the world's population, which breaks down to an average of 21 metric tons of carbon dioxide per person per year. By 2019, those four industrialized countries made up 18 percent of the world's emissions — not because they reduced emissions, but because the developing nations started catching up. The only one of those four big polluters that cut its emissions a lot was the U.K., down from 21 metric tons per person to 5.4. Emissions per capita from the United States, Canada, and Australia remained high — at 15.5 metric tons per person. Per person emissions from India went up to 2.47 metric tons per person, while China went up to 7.4 metric tons per person.

When trying to figure out who's actually causing the carbon dioxide emissions of a country such as China, consider that the industrialized world imports many of the goods that country manufactures. One glance at the label on your shirt or the fine print under your breakfast bowl may tell you that it was made in China. The government of Sweden has publicly recognized that part of the industry emissions in China are growing simply to provide industrialized countries with things they want or need.

GOOD NEWS

Advocates for climate protection definitely have high hopes for these developing countries. China has invested more money than any country in renewable energy. Brazil, India, and China plan to make cuts in their emissions over the next five years that, taken together, will exceed the United States' planned GHG reductions. The progressive work of these three countries — if it continues at this rate — is at the level needed to help the world level off GHG emissions in time to avoid irreversible changes.

China

China surpassed the United States as the world's top polluter in the spring of 2007 — an event that many analysts thought would take at least another few years. Despite this dubious achievement, China is taking steps to improve its environmental record. Without efforts to reduce emissions, including closing a large number of older, more polluting coal plants, China's emissions would have increased by more than 7 percent of actual increases. Still, China continues to increase pollution, with a worrying surge in 2021, after nearly a decade of slowing growth. Due to pandemic responses in other countries, the demand for Chinese goods for export actually increased through 2021. This growth in demand for goods from China caused an increase in energy demand.

China continues to build coal plants — although more efficient ones than the plants it closed. Some promising news is that the price of solar for much of China has dropped such that it's cheaper for China to decarbonize than continue to rely on coal.

One big driver for China to reduce reliance on coal has been the extreme pollution that shrouds big cities like Beijing. A study by the RAND Institute estimated that acute pollution costs China more than $535 billion per year in lost productivity. According to the Center for Strategic & International Studies (CSIS), a bipartisan, nonprofit policy research organization focused on advancing practical ideas to address the world's greatest challenges, China has shut down all the coal-powered stations surrounding its capital city — but it has a very long way to go to remove coal-fired electricity altogether. That's why China and India insisted at the Glasgow COP26 meeting that the words "phase out" coal, be changed to "phase down."

The following sections address some more specifics that China has done and has promised to do and some challenges facing the country.

Electric improvements

With its Renewable Energy Law and Energy Conservation Plan in place, China is cutting back on emissions. The Chinese government recently achieved a five-year goal of reducing GHGs from electricity production by 5 percent below their projected levels. This reduction had the same effect as shutting down more than 20 big coal-fired power plants.

China is also seeking to develop in a greener direction. It leads the world in its investments in renewable energy. In 2018 alone, China invested more than $90 billion (US dollars). The country has been using wind and solar power to provide electricity to remote villages across six regions. Because these areas are remote, connecting them to the power grid would be difficult. The alternative? Go off the grid and be self-sustaining. China's investments in solar energy has lowered the cost of solar energy so that it's now cheaper than coal. (See Chapter 13 for more about renewable energy.) The 13th Five Year Plan for Electricity (2016–2020) aimed to raise nonfossil fuel's share of total electricity production from 35 to 39 percent by 2020, per CSIS.

China committed to make nonfossil fuel energy 20 percent of its energy supply by 2030 and to peak carbon dioxide emissions by 2030. Chinese President Xi Jinping expanded on that commitment in a speech to the United Nations in September 2020 when he announced that China aims to achieve carbon neutrality by 2060. As is the case for the vast majority of nations, China hasn't done nearly enough, but every step matters. There is no solution to the climate crisis without China cutting its emissions dramatically.

China is the world's largest investor in clean energy. Between 2013 and 2018, the country's investments in renewables grew from $53.3 billion to an impressive peak of $125 billion. This figure has fallen in recent years, but in 2019 China's investments still stood at $83.4 billion — roughly 23 percent of global renewable energy investment. At COP26 it promised to boost wind and solar power capacity to more than 1,200 gigawatts, with nonfossil fuel primary energy production up to one quarter of energy.

China is also becoming the largest market in the world for renewable energy. Reports estimate that 1 in every 4 gigawatts of global renewable energy will be generated by China through 2040.

Changes on wheels

The Chinese government is responding to the country's changing transportation needs. Not long ago, a picture of a Chinese street showed a sea of bicycles. Everyone, it seemed, rode one. Today, that's changing. The *China Daily* newspaper reports that Beijing, China's acclaimed Bicycle Kingdom, now contains up to 2 million cars (although it still has 3 to 4 million bikes) because middle-class residents are changing how they get around.

To deal with this challenge, China's government implemented vehicle emission standards as part of a National Environmental Friendly Vehicles project that cut emissions by 5 percent below what they otherwise would have been by 2020 — the most rigorous standards in the world. China also taxes cars based on the size of the engine, which means that SUV and truck drivers pay more for their less-efficient vehicles.

With aggressive centralized policies, China is now both the largest manufacturer and the largest buyer of electric vehicles (EVs) in the world, accounting for more than half of all electric cars made and sold in the world in 2018. China also makes 99 percent of the world's electric buses.

The challenges

WARNING

Even with the steps in the right direction that we talk about in the preceding sections, the growth in GHGs in China is a globally worrying trend. The country continues to use coal and gasoline as its primary fuel sources — two fuels that produce the most GHG. Coal supplies 61 percent of China's energy, according to the International Energy Agency (IEA), per CSIS.

Chinese economic goals are on a collision course with global environmental goals. With China adopting goals to end poverty, and with a growing middle class chomping at the bit to join the lifestyle that rich industrialized countries have created, its emissions are bound to increase unless it shifts its development in a climate-friendly direction. China faces a tough challenge: reducing emissions while raising the economic status of its population. In addition, China is the largest exporter in the world of a number of GHG-intensive products. For example, China produces about 55 percent of the world's concrete, a very energy-intensive business.

Even though the climate change problem in very large in China, there seems to be a reasonable chance that with its centrally directed government the country may be able to reduce GHG emissions fast enough to help the world keep to 1.5 degrees C. The world needs to cheer China on — its success is vital for everyone everywhere.

Brazil

Brazil is caught in a tight spot. On one hand, it has one of the biologically richest ecosystems in the world, the Amazon rainforest, with the Amazon River second in length only to the Nile. The Amazon works as a rain machine, feeding the vast farmland crops that supply food to countries around the world. On the other hand, cutting down the Amazon rainforest to provide land for agriculture is reducing rainfall and creating 75 percent of the country's carbon dioxide emissions (from cutting down or burning trees).

Although the Brazilian government had been moving forward on some green-friendly initiatives, the election in 2019 of Jair Bolsonaro led to accelerated felling of the Amazon and more violence toward indigenous peoples. At the Glasgow Summit, COP26, Brazil joined many other countries in a pledge to end deforestation by 2030. Many observers are skeptical about Brazil's role in the alliance because the pledge isn't legally binding. Recent reports in 2022 show that, despite Bolsonaro's pledges at COP26, deforestation in the Amazon is accelerating.

CURITIBA: A LEADER IN PUBLIC TRANSPORTATION

One of the keys to reducing greenhouse gases is getting people out of their cars and onto public transit. The Brazilian city of Curitiba is a leader in this area. The concept for that city's system goes back to the early 1970s. The mayor, Jamie Lerner, was simply looking for a way to fix two problems — the poor couldn't afford to ride the buses, and the buses needed more passengers. He started a transit voucher system for people who collected recyclables (thus reducing litter) and instructed bus drivers to stop whenever someone waved, regardless of whether they were at a designated stop.

From this humble start, a much bigger dream emerged — to create a city that could integrate work, housing, and recreation to serve Curitiba's quickly growing population. Within three years of its startup, the public transit system served a third of the city's transportation needs. Over the years, the system has added buses and built subway lines, and the city has designed the system in a very efficient way — making sure that buses connect with subway stations and adding express bus lines.

Curitiba's plan has been so successful because it defined from the start how transportation would work in the city and, more importantly, how the city would develop while it continued to grow. It's one of the most substantial long-term transit plans done by a city anywhere in the world. In fact, reflecting on the success of Curitaba's Bus Rapid Transit (BRT), former mayor Lerner told *The Guardian*, "We started BRT in 1974; now 300 cities around the world are using it."

The next sections dive into the details of the bright spots of Brazil's climate journey — and its challenges.

Miles ahead (on ethanol)

Brazil has become the world leader in producing ethanol as part of its plan to wean the country off oil. South America's largest nation originally focused on ethanol to lower the country's dependence on foreign oil during the 1970 oil crisis — and it worked.

Ethanol production has grown year over year, even with a small downturn as demand increased for sugarcane to produce sugar instead of fuel. In 2019, 37 billion liters were produced, dropping to 26.72 billion liters in 2020. Still, Brazil's ethanol production is second only to production in the United States (ethanol in the U.S. is highly subsidized and is mostly made from corn, which is less productive that sugar cane and which reduces land available for food crops).

All the cars in Brazil now run on at least some ethanol. By law, all fuel has to be at least 20 percent and no more than 27 percent ethanol. As a result ethanol makes up half of the fuel sold in Brazil, according to the Earth Policy Institute. Initially Brazil's ethanol boom was entirely based on sugarcane, but in recent years, cheap and abundant corn has led to Brazil massively increasing both corn and sugarcane ethanol, according to the USDA. Still, 96 percent of Brazil's ethanol comes from sugarcane — more than 37 billion liters in 2019. In fact, a third of all sugarcane production in Brazil goes straight to producing this fuel. In 2020, Brazilian sugarcane contributed more than 17 percent of Brazil's total energy supply.

WARNING

Increasingly, researchers have linked ethanol production in Brazil to more deforestation — but these rumors are disputed. Some researchers, such as studies published in the *Journal for Sustainable Agronomy*, counter that sugar cane production has happened primarily in the southern region of the county, well away from the Amazon rainforest. Increasingly, however, ecologists worry that the economic success of sugarcane ethanol may lead to sugarcane replacing rainforest in the northern Amazon.

Jose Goldemberg, a scientist and former Brazilian cabinet minister who has long advocated ethanol, believes that not every country can use ethanol as a solution to reducing fossil fuel dependency; he maintains that ethanol is only one of many pieces needed to solve the climate change puzzle. Goldemberg has said that if even 10 percent of cars in the world were to run on ethanol, it would require ten times the amount of ethanol that Brazil currently produces — an unlikely possibility. (See the sidebar "The man behind the success of ethanol in Brazil," in this chapter, for the story of Goldemberg's contribution to ethanol production.)

THE MAN BEHIND THE SUCCESS OF ETHANOL IN BRAZIL

These days, you frequently see ethanol mentioned in the news headlines. But Jose Goldemberg has had ethanol on his mind for more than 40 years. He's known as the man behind the global success and acceptance of ethanol as a replacement fuel for gasoline.

Brazil felt a heavy dose of oil shock back in 1975, so, to save its economy, it launched a major campaign on alternative types of oil. Goldemberg was a nuclear physicist at the University of São Paulo at the time and published a paper in the journal *Science* three years after the oil shock. The topic? A message to the world that you can get ethanol from sugarcane — and that Brazil had created this clean and renewable alternative to conventional oil. His work provided the basis for today's ethanol hype.

In 1988, Goldemberg was one of the scientists who attended the landmark first comprehensive international scientific conference on the threat of global warming. The conference, held in Toronto, was entitled "Our Changing Atmosphere: Implications for Global Security." He assisted in drafting the consensus statement and developing a target for reductions 20 percent below 1988 levels as a first step by 2005. He attended the UN's 11th Conference of the Parties in Montreal in 2005, reporting on the agreement between the states of California and São Paulo.

Goldemberg has dedicated himself to energy issues his entire life. His academic work has led him to professorships at universities beyond Brazil — including Stanford, the University of Paris, and the University of Toronto. He published the acclaimed book *Energy for a Sustainable World* (Wiley) more than ten years ago, and at the age of 93, he continues to work on developing policy to solve the world's growing energy needs.

But heightened worldwide interest in ethanol makes Goldemberg certain that Brazil will boost production by 50 percent. In Brazil, the majority of cars on the road are *flexible-fuel* — they run on either gasoline or ethanol — and most people choose the more economical ethanol. The use of ethanol in Brazil has avoided 515 million tons of carbon, according to Renewables Now, an independent news site.

A cut below the rest

Deforestation is the biggest roadblock to reducing emissions for Brazil, accounting for 75 percent of the country's annual carbon emissions. While commodity prices for crops go up, farmers want to plant more crops and clear more space for cattle — so, they often remove forests. Seventeen percent of the original Amazonian rainforest has disappeared — an area larger than the size of France.

The rainforest of Brazil is large and difficult to manage — the trees cover an area the size of the whole western United States. No one claims ownership of much of the land, or people are disputing who owns it. So, the Brazilian government can't easily regulate forest activities. According to *Nature* Journal, Brazil, along with many other developing countries, has argued for including global assistance to stop illegal logging and deforestation.

India

In terms of its carbon emissions, India is a sleeping giant. According to the World Bank, the country has 17.7 percent of the world's population but currently accounts for only 7 percent of global carbon dioxide emissions. Even at 7 percent of world emissions, India is the third largest polluter. While the country becomes increasingly urbanized and industrialized, that number could skyrocket. Government agencies and organizations alike are taking the initiative to ensure that India starts reducing pollution. At the Glasgow climate meeting, COP26, Prime Minister Narendra Modi of India pledged to have fully half of India's energy from renewable sources by 2030. That will be a huge 550 gigawatts of power.

India's long-term Paris target is a weak promise to achieve net zero by 2070. However, seasoned climate negotiators noted that they never imagined India would make a commitment to phase out fossil fuels. The commitment was a significant step for this major polluter.

**GOOD
NEWS**

Since 2001, India has spent more than 2 percent of its GDP on responding to climate change. Measures taken under the Kyoto Protocol's Clean Development Mechanism (CDM; we talk about these projects in the section "Choosing Sustainable Development," later in this chapter) brought about a reduction of more than 27 million metric tons of carbon dioxide. India continues to pursue projects under the new trading mechanisms set out under the Paris Agreement that have replaced Kyoto's CDM (via the controversial Article 6 carbon trading system), according to the Energy Mix, a Canadian nonprofit that promotes community awareness of, engagement in, and action on climate change, energy, and the shift off carbon.

The next sections examine the details of what India has promised to the world community. Like China, it's a big deal that India has committed to reduce emissions at all — but the promises remain too little.

Improving energy efficiency

The Indian government has been working hard at improving the efficiency of nonrenewable power providers across the country since 2008 when it passed its first climate plan. For example, the government lowered coal subsidies. This loss of funding motivated coal plants to increase their efficiency and even replace some of the coal with natural gas.

India has pledged that by 2030 the internal combustion engine will be banned from new car sales.

Big business, big changes

Major businesses in India are getting involved in fighting global warming. And most big banks are coming on board by lending companies money to help cover the initial costs of energy-efficiency projects.

Some projects have shown an immediate payback:

>> Apollo Tyres Ltd., a tire production company, invested $22,500 in redoing their heating system, heating the building with the heat from the hot water system. With just a 14-month payback period, the company brings in an annual savings of $19,500. That success started more than a dozen years ago, and Apollo keeps up its path to sustainability.

>> Arvind Mills Ltd., also in India, is the largest producer of denim jeans in the world. They connected their two main cooling pumps so that they could turn one off in the winter, a project that saves $280,000 a year — and the project cost them nothing. With production in Ethiopia and India, the Ethiopian operations are 100 percent renewable and the Gujarat operations have the largest rooftop solar installation in the country.

A leader in renewable energy

India has become a leader in renewable energy (see Chapter 13). When this book was first published in 2009, renewable energy supplied 8 percent of the country's total energy needs. Today, an astonishing 38 percent of India's energy comes from renewable sources — and this number is growing. Under the Paris Agreement, India has promised to rely on renewable energy for 50 percent of its electricity by 2030.

India has invested heavily in wind and solar energy and is now the third top producer of renewable energy, just after China and the United States.

India also looks to another developing country for a profitable and carbon-friendly venture. The world's second-largest sugarcane producer after Brazil, India wants to follow in Brazil's footsteps and start producing ethanol. In 2003, India committed to shifting nine states to a gasoline blend that features 5 percent ethanol. By 2020, the goal posts have moved as India pledges to get to achieve a 20 percent blending target by 2025. Though no one knows the exact emission reduction that this measure will create, this move can open up both the Indian and the global market and lower the production price of ethanol.

Growing by the numbers

India's per capita emissions are very low — 2.47 Tco2e (tons of carbon dioxide equivalent), as compared to the global average of 6.45 tco2 per capita. Nevertheless, the country faces substantial pressures to reduce its emissions based on its large population and arguably unsustainable ways. Like China, India has a lot of coal. While India pursues economic prosperity, its growth in GHG emissions is a serious cause for global concern. India joined China at the Glasgow COP26 critical summit to weaken language about "phasing out" coal to "phasing down" coal. Still, India's commitment to phasing down coal is progress.

Choosing Sustainable Development

Although we can't overstate the seriousness of poverty and the poor living conditions that many people in developing nations face, these countries do have a tremendous opportunity: Developing nations have the chance to steer clear of the mistakes of the industrialized nations and develop in a way that doesn't harm the planet. Instead of building their nations' economies on carbon-dioxide–emitting fossil fuels, they can choose sustainable development.

Finding a way to eliminate global poverty, while eliminating the use of fossil fuels, is a challenge. The United Nations has taken on that challenge with something called the sustainable development goals (SDGs). Broken down into 17 topic areas, the SDG goals are endorsed by the governments of the world. The goals promote public health, literacy, economic sustainability, climate action, healthy oceans, and so much more. At the foundation of these goals is a concept called sustainable development, which the following sections discuss in greater detail.

Understanding what sustainable development is

What *sustainable development* means is open to debate. The 1987 Report of the World Commission on Environment and Development, Our Common Future (also known as the Brundtland Report), which launched the term's popularity, used several definitions. Here are the general concepts:

>> Development in both industrialized and developing countries that uses materials in an environmentally responsible way

>> Development that doesn't hurt the way that natural ecosystems function; doesn't endanger species; and avoids air, water, and soil pollution

>> Development that meets humanity's needs without using so many resources or harming ecosystems to the extent that future generations won't be able to meet their own needs

REMEMBER

Climate change and sustainable development are linked. A nation needs a strong and balanced economy (an aspect also in peril because of climate change) to affect sustainable development. Sustainable development promotes

>> **Renewable sources of energy:** These sustainable sources produce low or no emissions, so they don't add to the climate change problem.

>> **The health and well-being of people:** The affected people may be in jeopardy because of the effects of climate change. Many of them in urban areas are suffering from air pollution from burning coal.

Old economies have sacrificed the environment to achieve economic growth. The newly developing countries have a chance to end this historical connection — to develop, but to do so with an eye for the climate change consequences and the incoming effects of climate change.

Expressing what developing countries can do

Climate change presents a two-fold challenge to developing countries. While they develop, they need to *mitigate* (or lessen) the production of greenhouse gases. Secondly, they need to adapt to the effects of global warming that they're already feeling.

Mitigation

When it comes to mitigation, developing countries can *leapfrog* — literally, skip over — the traditional fossil-fuel–based model that the older industrial economies followed and move straight into renewable energies to avoid boosting GHGs. They can choose new ways of generating energy, such as using solar, wind, low-flow tidal, and geothermal power as well as biofuel if from sustainable sources. (We talk about these alternate energy sources in Chapter 13.) Developing countries can't always easily get these cleaner, renewable technologies, but the technologies have numerous benefits. Improvements in the energy sector also improve community health and general productivity, and boost the economy.

Adaptation

Although developing nations have the chance to avoid the mistakes made by industrialized countries, they can't avoid the consequences of the mistakes already made. Developing nations have been feeling the effects of global warming for decades. The impacts of climate change are likely to worsen, so developing nations must adapt to prevent avoidable loss of life and property. (We discuss the disproportionate effects of climate change on developing countries in Chapter 9.)

REMEMBER

Adaptation takes many forms:

>> Shifting to more drought-resistant crops

>> Rethinking transportation infrastructure to locate bridges away from areas vulnerable to flash floods

>> Protecting and rehabilitating mangrove forests, which can help protect coastlines from increased storm surges

>> Moving vulnerable populations away from low-lying islands or coastal zones that may become flooded out of existence

Some countries need to take all these measures — and more.

One major adaptation project that developing countries can undertake is planting trees in areas suffering from deforestation. Planting trees not only cools down the planet (a positive characteristic of trees and plants that we talk about in Chapter 2), but it also enriches the soil with nutrients and helps reduce water runoff from flashfloods. The soil on a treeless hillside washes away in a mudslide, but a tree-covered hillside's soil stays put. The fewer natural disasters in a developing country, the less damage the local community and economy suffer.

Sadly, adapting to global warming also means preparing for the worst. Developing countries need safe water and food supplies to ensure that people have enough to survive if one of the extreme weather events — such as hurricanes and floods — wallops the developing world. To guard against those storms, governments need to build higher dikes and make stronger bridges. They need to make riverbanks and seasides capable of coping with deluges. They also need to make plans for a world with too little water; already, the World Bank is working with many other development agencies around the world, such as its work with several Caribbean countries to develop drought-resistant crops, while helping Bangladesh plan ahead for serious loss of territory due to sea level rise.

REPLANTING MANGROVES IS SOME GOOD NEWS

The mangrove forests of coastal zones near the equator are one of the most endangered ecosystems in the world — even more endangered than rainforests. These forests that are rooted in ocean saltwater, sometimes dry and sometimes wet. Mangroves are astonishingly effective in keeping carbon out of the atmosphere — up to 50 times more effective than other ecosystems. Mangroves are also rich in biodiversity, providing habitat for fisheries, home for the rare proboscis monkey, flamingos, sharks, manatees, and a multitude of other species.

Furthermore, mangroves provide protection for coastal settlements against storms and even tsunamis. Yet, this very valuable ecosystem is under assault from human activities — with mangroves removed to create tourism destinations and to farm shrimp in artificial ponds where mangroves used to be. Globally, from 1980 to 2005, as many as 3 million hectares of mangroves were cut down. The damage continues as mangroves have been disappearing twice as fast as rainforests.

Mangroves are finally getting the respect they deserve. Critical science and research has found the best approach to replant mangroves. Mangrove forests once replanted don't run the same risk as land-based forests of burning down.

All around the world, mangroves are being replanted, according to Rethink, a news source that stands for resilience thinking for global development. In Vietnam, Sri Lanka, Honduras, Guyana, Mozambique, and the Solomon Islands — as well as the U.S. state of Florida — mangroves are being restored. The best approach is often to let nature take over and let the mangroves restore themselves by creating the right conditions. Restoring mangrove forests provides real hope.

Adaptation also means adjusting to whatever changes climate brings. Some communities may need to relocate if they can no longer sustain themselves. If a flood washes out a road or a hurricane levels a town, you shouldn't rebuild in the same way or in the same locations.

How industrialized countries can help

Industrialized countries have three roles to play in helping developing countries adapt to climate change — as leaders, funders, and partners:

Leaders

Industrialized countries need to take the lead in clean technologies to show developing countries that the industrialized world is dedicated to cutting greenhouse gas emissions and that it recognizes its part in creating most of the emissions to date. (We talk about the actions that wealthy nations can take — both cutting emissions and adapting to climate change — in Chapter 11.)

Funders

In 2009 at the Copenhagen COP15, the United States made a pledge that by 2020, wealthy countries, export development agencies, and private sector donors, would ensure that a Climate Fund would have $100 U.S. billion per year to assist developing countries. Developing countries didn't celebrate this pledge. They made it clear that what they wanted was wealthy countries to commit to slash emissions. As the lead negotiator for island nation of Tuvulu stated, "We do not want your thirty pieces of silver . . . . Our future is not for sale."

ENDANGERED NATIONS

In 1987, at the General Assembly of the United Nations, President Maumoon Abdul Gayoon of the Maldives warned that the world would soon have to use a new term. Humanity would talk about not just endangered species, but endangered nations. He urged that the world's wealthiest nations act on global warming.

Ten years later, President Gayoon spoke at the United Nations to mark five years since the Rio Earth Summit, saying that the world had ignored the pleas of low-lying island states. As a result, the Maldives was forced to relocate villagers, as well as the population of an entire island, to higher ground.

Over the years, under different political leaders, the Maldives have continued to press industrialized countries to slash their emissions, to contribute needed financing to developing countries, and to help with adaptation. The Maldivian government knows that if the world warms to 3.6 degrees F (2 degrees C), their whole nation will disappear. The saying "1.5 to stay alive" has real significance in low-lying island countries like the Maldives.

To make this visible in dramatic fashion, on October 17, 2009, President Mohammed Nasheed and 13 government officials held their cabinet meeting — underwater in scuba gear!

The wealthy nations supported the U.S. initiative, and it became a part of subsequent climate negotiations. Unfortunately, at COP26 in 2021, the wealthy countries admitted they hadn't delivered, but they made a new promise with a road map to deliver by 2023.

Partners

Beyond developing and sharing low-carbon technologies, the United Nations Framework Convention on Climate Change (UNFCCC) says that industrialized countries need to partner with developing countries to help them cope with climate change. Industrialized countries have both the resources and the responsibility to work on these projects.

TECHNICAL
STUFF

Under the Paris Agreement, mechanisms developed under the Kyoto Protocol have been replaced. Wealthy countries are to assist developing countries in the following ways:

CLIMATE FINANCE THROUGH THE $100 U.S. BILLION PER YEAR

Climate finance and the $100 billion is pretty straightforward. The wealthy countries promised and now have to deliver. Those funds can be used for leapfrogging fossil fuels and finding better, more effective ways to deliver energy.

ADAPTATION FINANCING THROUGH THE SEPARATE ADAPTATION FUND

Adaptation funding is also straightforward but has been less well attended to. Developing countries are already experiencing massive losses in life and property due to climate change–related events. They need funds to protect key infrastructure from extreme weather events, whether roads, bridges, and coastlines — or to revamp agricultural practices to be drought resistant. Adaptation is desperately needed — and not just in poor countries.

Rich countries also have ignored the need to adapt. People living in British Columbia discovered that in the summer of 2021 when nearly 600 people died in a *heat dome* — when high pressure stays over the same area for days or even weeks, trapping extremely warm air underneath. Through the same period, 1,600 wildfires burned out of control, and a few months later floods and landslides created chaos, loss of lives, and property destruction. Canada has resources for recovery, but southern India, hit with similar disasters, doesn't.

ARTICLE 6 AND CARBON TRADING

According to Ecosystem Marketplace, Article 6 will allow for carbon trading, but it also has the goal of meeting needs for mitigation and adaptation through the delivery of finance, technology transfer, knowledge sharing, and capacity-building. The conversations about trading pollution has changed since 1997 and Kyoto. The early trades under Kyoto's' CDM turned out to be less effective than was hoped.

The idea was that wealthy countries would get credit for reducing GHGs that were actually reduced in a developing country. The wealthy country would give the poorer country cash to reduce emissions, and then count the emission reductions against its own target. The theory was that because the atmosphere doesn't care where the reductions come from — the atmosphere just needs countries to massively cut the dumping of warming gases.

But CDM soon showed it had real flaws. Not always, but sometimes, scams were possible. A developing country might have been about to cut emissions anyway. So it shouldn't have been treated as a new win. That problem gets called *additionality*. A developing country might have been about to cut emissions anyway, so the reduction in emissions shouldn't have counted as something new. In other words, not additional. And sometimes the emission reductions weren't verified (see Chapter 10 on CDM and emissions trading).

Article 6 has been developed to avoid those flaws while encouraging partnerships.

LOSS AND DAMAGE

Also known as the Warsaw Mechanism for Loss and Damage, loss and damage (L&D) is the most contentious funding mechanism is. This kind of funding in principle is a compensation fund for damage and losses that occur due to the climate crisis, recognizing that wealthy nations primarily caused climate crisis.

It emerged as a hot topic in the Warsaw COP19 in 2013. As the conference started, Typhoon Haiyan had just hit the Philippines, killing 6,000 people and displacing more than 4 million. Young Filipino negotiator Yeb Sano hadn't been able to reach his family. He feared his father might have died. In desperation to get the world's attention, he went on a hunger strike at the Warsaw conference. Many others joined in for the two week COP (including your co-author Elizabeth!) to draw attention to the desperate situation. The result was the Warsaw Mechanism on Loss and Damage. So far, only Scotland has made a donation to it. The United States and Canada fear the program could create a legal liability — even though the text of the Paris Agreement says it doesn't.

In the end, Sano got home to find his own family had survived — but so too did the demand from developing countries for L&D funds to assist when extreme weather events linked to climate change hits hard.

PLANNING ON IT

The world can no longer avoid some changes caused by global warming. All countries need to come up with plans for how to deal with and adapt to climate change — a plan that countries actually use and update on a regular basis. As the climate changes, so must the plans and strategies of countries around the world.

Making adaptation plans for climate change, which is so diverse in its effects, means dedicated cooperation between government departments and across sectors of government, industry, business, and the public. Regular communication between different regions helps coordinate the planning and enables the government to make adjustments to the plan based on successes and failures.

For developing countries, plans for adaptation must include stable funding. This funding can come directly from industrialized countries or from organizations funded by those countries — such as the Global Environment Fund (GEF). The GEF deals with a range of environmental issues, including climate change. It gives grants to projects in developing countries around the world.

Unfortunately, the world has made fewer meaningful efforts on adaptation, particularly for the developing world, than its efforts to reduce GHGs. "Too little, too late" isn't the epitaph humanity wants! The world has to deal with climate change right now, and every nation needs to plan for today, tomorrow, and well into the future to literally live with the changes.

5

Solving the Problem

Discover the astonishing rising use and falling costs of renewable energy sources, such as wind and solar power.

Take a look at how rapidly changes in business and infrastructure can happen when old ways are disrupted by new ones.

Check out what nongovernmental organizations are doing to hold governments' and industries' feet to their own fire.

Understand how the media is affecting human's understanding of climate change.

Determine whether there's enough time left to make the vital changes to energy, industry, and consumption.

Chapter **13**

Powering the World — Renewable Green Energy

O il and other fossil fuels have been the world's major sources of energy since the Industrial Revolution. And using that fossil energy has caused and keeps on causing global warming and other changes, but times are changing. Global average temperature has been rising since fossil fuels (coal, first) started to be used around 1850. The global average temperature is now 1.4 degrees F (1.1 degrees C) above pre-industrial times. That may not sound like a lot, but remember, a change in the world's global average temperature is a big change. At 1.4 degrees F (1.1 degree C) more warming, the effects are everywhere: unprecedented droughts, fires, floods; enormous storms; damaged property and infrastructure; and loss of life. Humanity has changed the chemistry of the global atmosphere.

Based on the greenhouse gases (GHGs) already emitted, exerting a warming influence on Earth, climate scientists now agree there's no stopping at 1.4 degrees F (1.1 degrees C). Without huge effort, the planet is headed for ever higher temperatures. At this writing, the Earth is on track for close to 5.4 degrees F (3 degrees C) warming. The world's leaders all agree that the use of fossil fuels must be reduced dramatically and quickly.

REMEMBER

Burning fossil fuels produces atmospheric carbon that results in global warming. The science says that to keep to a maximum of 3.6 degrees F (1.5 degrees Celsius), carbon dioxide emissions must be cut by 45 percent below 2010 levels by 2030, and be reduced to net zero by 2050. *Net zero* means completely negating the amount of

GHGs produced by human activity, to be achieved by reducing emissions and implementing methods of absorbing carbon dioxide from the atmosphere.

Rapid decarbonization is required well before 2050. *Decarbonize* may be an unfamiliar verb, but it's one you'll hear more. Societies decarbonize when all the electricity used comes from nonfossil fuel sources, when cars aren't propelled by the internal combustion engine (ICE), and when people stop looking at oil, gas, and coal as sources of wealth, but as a threat to a livable world. Scientists have warned governments that humanity must achieve a huge amount of decarbonization before 2030.

You're probably in agreement that renewable energy is a secure energy source, helps reduce the use of fossil fuels, and gives people a real option for cutting GHG emissions. The International Energy Agency (IEA) projects that the global use of wind, solar, geothermal, and hydroelectric power will continue to grow exponentially. Oil and gas, except the very sweetest crude oil and locally produced natural gas, will eventually be drastically reduced.

The late Dr. Hermann Scheer, architect of Germany's Renewable Sources Act and the leading proponent for the creation of a United Nations–sanctioned International Renewable Energy Agency (IRENA), argued that not only is a significant shift toward 100 percent renewable energy is possible, but it's absolutely necessary.

In this chapter, we look at how the world's use of oil and other fossil fuels needs to change, and we explore the opportunities for renewable energy.

Addressing Energy Demand

Cutting the use of fossil fuels will reduce emissions and so help to limit climate change. But the world still needs energy, and lots of it. Economic and population growth create more demand for energy services such as mobility, heating, cooling, and lighting, which pushes up energy use. So demand for energy is going up, not down; especially in developing economies, demand will increase dramatically over the next century. These sections look at what can be done to meet this demand.

Being more efficient

Demand can be reduced a lot by using energy more efficiently. Using energy efficiently essentially means doing more with less. Over the last century, economies have massively improved labor productivity — work done by human labor. What was once the work of hundreds is now done by a handful of workers. And a lot of that was due to using fossil fuels instead of human labor.

Energy analysts believe the world's economies have the potential to increase the productivity of energy just as remarkably as what happened with the productivity of labor — as much as by a factor of ten. Ten times as much work done — whether in how people move around or how they keep the lights on — from one unit of energy. Energy-efficiency improvements deliver significant benefits for the climate, national budgets, and energy consumers. For example, efficiency savings reduce energy use and expenditure, which improves affordability, particularly for households.

According to the IEA, "*energy intensity* is the amount of energy used to generate a unit of GDP. Remarkable progress is being made on this front: energy decreased globally by 36 percent between 1990 and 2018. In China, intensity more than halved (–70 percent) over this period."

The efficiency gains since 2000 in IEA member countries resulted in the avoidance of more than $600 U.S. billion of expenditure for fuels for heating, road transport, and a range of other energy end uses.

Just about anything — whether it's a whole office building, an entire fleet of cars, or a total manufacturing system — can be made more efficient. The obvious candidates for such dramatic efficiency increases are in the following areas:

>> **Buildings:** Technological developments in buildings have created more efficient insulation, better lighting systems, and better heating and cooling (a simple heat pump, for example, can reduce the energy demand of a residence heated by natural gas).

>> **Transportation:** Using electric vehicles (EVs) use less energy overall because they are lighter and require far less maintenance — they're much more efficient in using the energy required to move them. The same can be said of heavy vehicles using electricity or hydrogen — they're more efficient from the outset, so in addition to reducing GHG emissions, they simply use less energy overall.

>> **Industry:** Industry is always looking for lower-cost and more-efficient ways to operate — the highest GHG emissions in modern industry are from the use of coal to make steel, from the carbon dioxide generated in making concrete, and, ironically, in the production of fossil fuels themselves (about 12 percent of Canada's total GHG emissions come from the use of natural gas in oil sands mining). There are two ways GHG from steel and concrete are already being reduced.

As clean electricity from renewable sources is developed, these three areas are expected to be dramatically reduced. Furthermore, new technologies are emerging to change how concrete and steel are made.

Combining heat and power

When any fuel is burned to generate electricity, only about one-third of the energy actually generates any power. The other two-thirds is wasted. A lot of it goes up the chimney. This type of energy waste is more common than you might think. With regular light bulbs, only 10 percent of the energy creates light — the other 90 percent is lost as heat! LEDs make no heat at all and use 85 percent less electricity than regular bulbs (refer to the nearby sidebar about light bulbs).

This lost energy doesn't have to go up in smoke; it can be used for co-generation, producing both heat and electricity at the same time. *Co-generation* takes the same amount of fuel, but it captures the heat that would be lost otherwise and uses it to heat a building. Figure 13-1 shows a co-generation system that uses turbines to reuse energy from pressurized blast furnace gas.

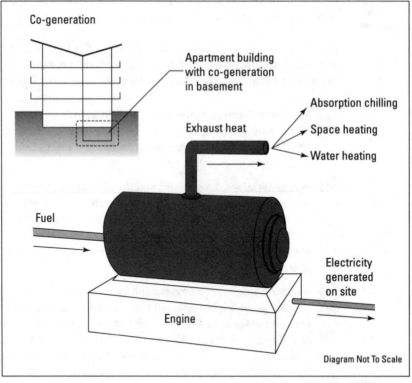

Co-generation

Apartment building
with co-generation
in basement

Absorption chilling

Exhaust heat

Space heating

Water heating

Fuel

Electricity
generated
on site

Engine

Diagram Not To Scale

© John Wiley & Sons, Inc.

FIGURE 13-1: A co-generation system takes wasted heat from a normal system and turns it into heat that a building can use.

HOW MUCH CARBON DOES IT TAKE TO CHANGE A LIGHT BULB?

When you flip a light switch, do you care what kind of bulb lights your room? Probably not. But the type of light bulb you choose to use is an easy area to make big energy efficiency gains. Consider the following types of bulbs:

- **Incandescent:** This is the old-fashioned bulb — the kind of bulb in cartoons that pops over someone's head when they get lit up by an idea. However, those types of bulbs aren't so smart anymore. In fact, incandescent bulbs waste 90 percent of energy used! Each incandescent bulb emits 4,405 pounds of carbon dioxide per year.

- **Compact fluorescent light bulbs (CFLs):** CFLs were the first energy saver competitor to incandescent bulbs. CFLs are the bulbs with a twisty spiral of tubing. They made big improvements in energy efficiency, creating much less warming gas — 1,028 pounds of carbon dioxide a year. Although CFLs had a very long lifetime, they did have mercury fumes inside the glass, which required careful disposal.

- **Light emitting diode bulbs (LEDs):** LEDs are the clear winner is the latest improvement. The LED gets to a low 514 pounds of carbon dioxide per year. LEDs create a nice warm light and can be ramped up for big lighting needs, like street lights.

As the use of gas is phased out, co-generation technology can also be used with renewable power plants that use biomass (like forest waste) or biogas (produced by using bacteria to eat up cow manure and food wastes) and can be supplemented by solar hot water heaters that pre-heat the water used to produce steam in the generators, meaning less biofuel needs to be used.

Changing How to Handle Fossil Fuel's Emissions

Even when civilization uses fewer fossil fuels and improves how efficiently it uses them, fossil fuels will still produce GHGs when burned. Currently, those gases are released into the atmosphere, but oil companies and researchers are exploring alternatives.

These new technologies are controversial. As the global carbon budget shrinks, the best investments are those that get the Earth off fossil fuels altogether. However, the fossil fuel industry wants to find a lifeline to keep producing its products.

Attentive climate-aware citizens will want to be sure the industry is using its own money to develop these innovative approaches. Overwhelmingly over the last few years, governments, such as Australia, Norway, Canada, and the United States have taken money allocated to increase use of renewable and green energy — like wind and solar — and allocated it for the following types of technologies that could allow fossil fuels to expand.

REMEMBER

The IPCC's estimates of how the world reaches net zero by 2050 include assumptions that as yet unproven technologies to store carbon will actually work. The IPCC includes carbon capture and storage (CCS — refer to the nearby sidebar for more information) as part of a broader category of *carbon dioxide removal (CDR)*. In all the IPCC pathways consistent with 3.6 degrees F (1.5 degrees C), CDR is used to neutralize emissions from sources for which there are no easy mitigation measures. The IPCC points out that "CDR deployed at scale is unproven, and reliance on such technology is a major risk in the ability to limit warming to 1.5. CDR is needed less in pathways with strong emphasis on energy efficiency and reduce demand."

The next sections look at the promise and the greenwashing of new technologies for humanity having our cake and eating it too.

Capturing and storing carbon dioxide

Oil companies and researchers are conducting major tests dealing with the capture and underground storage of GHG. Storing the carbon works differently in different places. Two ways of capturing and storing carbon are currently being developed include the following:

>> **During oil production:** Oil-product producers pump the carbon dioxide into the ground at the same time that they pump the oil out of the ground. The pressure from the gas being pumped in actually helps to get the oil out more efficiently — creating a controversy over whether this use of the technology actually benefits the climate. (See the nearby sidebar for more information.)

>> **Dealing with industrial emissions:** Major carbon dioxide emitters, such as coal power plants, capture the gas by containing the source of the emissions and directing the gas underground. This process stores the carbon dioxide in places from which people once extracted oil; in big, empty spaces underground; or into unmineable coal beds and saltwater aquifers. This need for storage-space limits the widespread use of this technology.

Figure 13-2 shows how both of these methods are supposed to work.

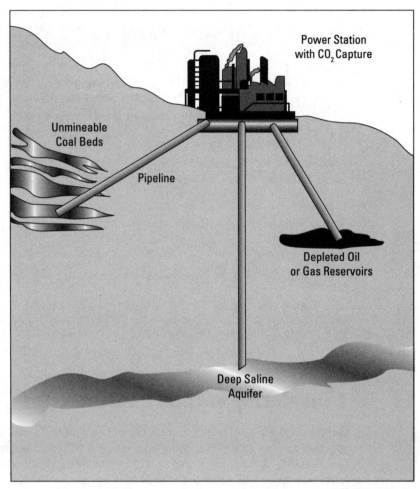

Power Station
with CO_2 Capture

Unmineable
Coal Beds

Pipeline

Depleted Oil
or Gas Reservoirs

Deep Saline
Aquifer

FIGURE 13-2:
How carbon
capture and
storage happens
in theory.

An area's geology is a huge factor in determining whether or not carbon capture and storage will work there. In Norway, for example, the technology is being applied by pumping carbon dioxide down under the deep sea. But what works in the North Sea geologically may not work elsewhere.

People can't capture all carbon dioxide emissions. Catching the emissions coming out of a stationary smoke stack is much easier than capturing the carbon blowing out of the tailpipe of a moving car. More than half of the carbon dioxide that can be captured is emitted by coal power plants; the remainder is from major industrial emissions.

CARBON CAPTURE CONTROVERSY

In Canada, the Weyburn Saskatchewan Project was touted as having real potential to capture the emissions from the Great Plains plant in Beula, North Dakota. The carbon dioxide emissions are transported through a 202-mile (325-km) pipeline and stored in old oil fields at Weyburn. This project has so far failed to meet any of its objectives. It was supposed to store up to 20 million metric tons of carbon dioxide — while producing another 130 million barrels of oil. The oil produced comes from injecting the carbon dioxide down the drilling well to help dislodge oil that wasn't exploited in the first phases of the well. In 2021, the Weyburn project stored only 2 million tons.

So too have proposed plants in the United States failed. In 2003 a big project, with more than a billion dollars in state and federal government support, was announced for Illinois. By 2007 the project stalled.

Norway and Australia similarly have a list of failed projects after wasting billions.

Some argue that they wouldn't be able to access that much oil without the carbon dioxide; so, the entire process may ultimately produce more carbon dioxide than if none had been captured at all.

Considering carbon capture cons

Carbon dioxide capture and storage depends on suitable and available storage space. According to the IEA, the planet has enough room underground to store tens to hundreds of years' worth of carbon dioxide emissions (no one knows just how much room exists, hence the wide time range).

Whether people can pipe carbon dioxide back underground, down the now-empty oil wells, depends on the kind of geology near the pollution source. The final reports from the Weyburn project indicate that in this particular rock formation carbon capture is technically feasible. Carbon capture and storage (CCS) is also in use in the North Sea. But technical feasibility on a small scale doesn't mean that it can be scaled to meet the very large projections of CCS proponents and is very different from being economically successful. Much remains to be demonstrated.

WARNING

Another huge issue is that the stored gas leaks. Escaped carbon dioxide from CCS underground is dangerous in two ways: Of course, released GHGs are dangerous for the climate, but also because carbon dioxide is colorless, odorless, and totally toxic, it could be lethal. Of course, humans breathe carbon dioxide every day — but remember — even though 412 ppm carbon dioxide is more than any human has ever breathed, the carbon leakage from a CCS facility would be 100 percent carbon dioxide. Carbon dioxide is heavier than air and would pool in low-lying areas and could kill people without warning.

WARNING

Some environmental groups have major concerns about CCS, arguing that it's promoted mainly as a delay tactic because it

>> Allows civilization to keep using fossil fuels, rather than replacing them with carbon-friendly alternatives.

>> Enables oil producers to extract more oil, ultimately creating more carbon dioxide emissions (see the nearby sidebar for more).

Despite challenges and concerns, CCS is in use in industrialized countries now and is hoped to play a big role in reducing worldwide GHG emissions by 2050. But note again that CCS is mainly needed to reduce GHG emissions from fossil fuel production (the production of concrete and steel are other majors) — so, if fossil fuel production itself is limited, then the need for CCS is reduced.

Reducing Energy Demand

Changing habits and technology is no easy feat, but it's something everyone will be doing in the coming years. The world has begun to think of fossil fuels as valuable but dangerous resources that must be managed carefully. Modern civilization's love affair with oil has been given a hard deadline by the most recent IPCC and other work. A 45 percent reduction in carbon dioxide emissions against 2010 levels by 2030 means no new development of fossil resources and cutting the existing use of fossil fuels wherever possible.

The International Energy Agency (IEA) has presented three scenarios for world energy demand from now until 2030:

>> **Stated policies:** This scenario (this is what nations are committed to today) calls for increases in renewable energy and increases in oil, natural gas, and nuclear energy and a reduction in the use of coal.

>> **Sustainable developments:** In this scenario, renewables rise a lot and nuclear comes up slightly, gas remains the same (neither up nor down), and the other fossils decrease significantly.

>> **Net zero by 2050:** In this scenario, all the fossils are cut way back, and renewables and nuclear take up the slack.

This data isn't an easy pill for modern economies to swallow. Of course, existing industries are resistant, and governments everywhere like to delay the inevitable so that the political effects fall on someone else's watch. But it's now beyond doubt that humanity needs to start using oil only where it's really needed and replacing it with alternatives where they exist. The good news is that the development of other energy sources and technologies has been so rapid that replacement is now feasible. The remaining question is how long will it take?

Fossil fuels will continue as an energy source in the interim in two areas:

>> In locations like the far north where solar and renewables *can't* deliver enough reliable energy to be counted on. Diesel will probably still be the go-to energy source until sufficient sources of biodiesel are developed.

>> Natural gas may be the optimal backup for solar and wind until battery technology and other storage options mature.

Developing the alternate biofuel, hydrogen, and ammonia energy sources that will eventually replace almost all fossil fuels will take time. (We discuss these and other fuel alternatives in next section.) Alternative forms of car fuel and energy to power buildings and industry have been around for a long time — ranging from electric cars to renewable energy sources. Even some alternatives to consumer goods are readily available. Plastics can be made from corn oil. Fleece and other synthetic materials can be made from recycled plastic bottles. Other alternatives are in the planning stages: One example involves a possible solar-powered airplane — the sky's the limit!

Investigating Renewable Energy Options

Sustainable, or *renewable*, energy sources — such as plants, the wind, and the sun — regenerate themselves or can be replenished. Once up and working, most of these energy sources produce zero or little GHG. The only emissions from wind and solar power, for example, could come from the manufacturing of the turbines and panels, not from the actual production of electricity.

REMEMBER

None of these sources can supply all the world's energy needs on its own, but collectively, renewables can replace fossil fuels. The solution lies in choosing which energy source is right for a particular region. For a secure and sustainable future, humanity needs a range of energy sources, making civilization's energy systems less vulnerable to changes — like having a bunch of backup systems.

TECHNICAL STUFF

The International Renewable Energy Agency (IRENA) says that renewable energy sources make up 11.5 percent of the energy used around the world. It's 8.5 percent in Australia, 27 percent in Canada, 15 percent in the United Kingdom, and 8.7 percent in the United States.

The following is a breakdown of renewable energy in 2020:

- **Hydropower:** 39.5 percent

- **Solar photovoltaic (PV):** 24.3 percent

- **Onshore wind:** 23.9 percent

- **Pumped storage:** 4.1 percent

- **Solid biofuels:** 3 percent

- Others, including mixed hydro, offshore wind, biogas, liquid biofuels, geothermal, marine, and renewable municipal waste account for the rest.

WARNING

About half comes from large-scale hydroelectric systems, although calling big hydro renewable and sustainable is controversial because they can permanently inundate thousands of square miles of territory for enormous reservoirs. Refer to the section, "Hydropower," later in this chapter for more information.

For all of the renewables in this list, except for large hydroelectric, the efficiency of generating usable energy is increasing, and costs are quickly dropping quickly. Specifically the costs for large scale solar PV are falling rapidly, for rooftop solar are decreasing but not as quickly, and for on- and off-shore wind are going down but not as fast.

The following sections discuss the main options available for producing usable energy from renewable sources.

Blowin' in the wind

With the power to uproot trees, tear apart homes, and transport Dorothy to Oz, the wind is a natural force to be reckoned with. Humans can harness that power to generate energy through overgrown pinwheels, known as wind turbines. A *wind turbine* typically features a propeller that has three blades on a hub that sits atop a tall pole. When the wind blows, the blades of the propeller spin, capturing the wind's power and transforming it into electricity or mechanical energy. Together, multiple turbines are called a *wind farm*.

Wind energy has a significant upside: It produces zero emissions. The IEA expects that building up the wind industry can help make huge reductions in GHG emissions. Worldwide, the wind-power industry has been growing at 23 to 25 percent a year since the early 1970s. And, while the industry grows, wind power technology has become more reliable and cheaper.

To get a sense of how fast things are changing, consider that in 2008, the total wind power capacity worldwide was about 8 percent of total renewables, concentrated in Denmark, Germany, and Spain. In 2020, it was about 25 percent, with China, the European Union, and the United States leading, with China's deployment of wind energy is growing the fastest.

Unfortunately, a few barriers have been slowing the implementation of wind technology, although these are becoming less prohibitive as the technology and distribution systems matures. These barriers include the following:

>> **Costly to connect:** Good sites for wind power are often far from transmission lines, and installing new lines is hard and costly. This difficulty is easing as many countries have realized the benefits of modernizing their electrical grids.

>> **No storage:** Wind power is available only when the wind blows. This difficulty also arises with solar power and sunshine, so similar solutions arise. Denmark, for example, sends the electricity generated by its wind power to Norway, where it's used to pump water to higher-elevation dams. When the wind stops, the water is released again through turbines to generate electricity that is sent back to Denmark. Making the connected grid larger makes this uneven power output less of a problem.

>> **Public opposition:** Public acceptance can be low, depending on where builders place onshore wind farms. People don't often want to have wind farms in their communities because they think those farms are unsightly. Wind farms also produce low-level noise frequencies that some people find unpleasant. Finally, some people worry that wind farms endanger birds. (For more, see the nearby sidebar.) Most large-scale wind installations are now offshore, easing acceptance.

A BIG FLAP OVER WIND POWER

One of the biggest complaints against wind turbines is that they kill birds. Initially, this threat was a serious problem because builders didn't take bird and bat migration routes into consideration when erecting the turbines. A report put out by the U.S. National Academy of Sciences says that for every 30 wind turbines, one bird is killed each year, although these numbers are contested.

As wind energy moves forward, wind turbine developers and companies are taking concerns for bird safety into account. They usually build turbines away from migration routes and often add colored markings to the blades to ensure that birds can see them. Ultimately, wind turbines are far from a bird's worst enemy, making up just 0.003 percent of human-caused bird deaths in the United States. Cats, on the other hand, annually kill an estimated 2.4 billion birds around the world, according to the American Bird Conservancy. Perhaps wind turbines aren't birds' biggest problem?

Here comes the sun

Solar energy is very efficient and, like wind power, it produces no GHGs. You can capture solar energy, even on cloudy days. And it's the most abundant type of energy available — far beyond civilization's needs. In fact, the World Energy Council reports that the amount of solar energy that reaches the earth is more than 7,500 times that of civilization's current annual energy demand! Even if society only captured 0.1 percent of the sun's energy, and used it with just 10 percent efficiency, it could still meet its current energy demand four times over.

People can harness solar energy in three ways, each of which we investigate in the following sections.

Photovoltaic energy

When people think of solar power, they probably think of *photovoltaic* power sources, which use solar cells to convert sunlight into electricity (*photo* means light, and *voltaic* means voltage). You can use these solar cells on their own — to power things such as calculators, satellites, or flashing electric construction signs — or as a part of a power grid, in which they contribute to an area's energy source.

One of the most common ways to use this type of solar energy is on roofs — using solar panels, solar shingles, solar tiles, or even solar glazing on skylight windows. You can have them designed for any size building with varying energy needs. When Bill Clinton was president, he set a goal of a "Million Solar Roofs." Even though later president didn't support it, this effort helped the photovoltaic industry get off the ground in that country. By Earth Day 2021, 4 percent of homes in the United States were powered by solar energy — still just scratching the surface of the potential for rooftop solar.

As the use of photovoltaic power spreads, the price continues to comes down. Japan, Germany, and the United States lead the world in widespread solar cell use. Solar PV with battery backup is now less expensive than new fossil fuel–generating capacity. In some new large installations in very sunny desert locations, the price of the generated power is contracted to be less than one tenth of new hydroelectric plants (see Chapter 17).

Solar photovoltaic energy tends to produce the most power when the demand is highest (on hot days in the summer when people pump up their air conditioning). Because electricity demand is higher at these peak times, the price of electricity is very high, making solar power very competitive and useful — especially if it's being produced close to where it's being used (like on your roof) so that as little precious power is wasted as possible (transporting power over the lines uses up power).

Passive solar energy

You can use the sun as an energy source, even without special technology, simply by using its direct heat and light. This type of solar energy is called *passive* solar energy because you don't need to physically transform the energy to use it. For example, the heat from the sun coming through your kitchen window warms the room. (Cats are famous for taking advantage of passive solar energy.)

This technique works because it follows the sun. The sun rises in the east and sets in the west. It's also higher in the sky in the summer and lower in the winter. Using passive solar energy means shading yourself from the summer sun and bringing in the warmth of the winter sun.

For example, to take advantage of passive solar energy in your home, you can install most of your windows so they face south (north if you're in the southern hemisphere) and fully insulate the opposite side of the house. You can also put your most-used rooms, such as the kitchen and living room, on the south side of the house and put your bedrooms on the opposite side. With this kind of house design, the sun's warmth heats and lights the kitchen and living room during the day. This warmth eventually heats the whole house. At night, your well-insulated bedrooms keep in the heat that built up over the day. In the summer, the sun's rays from above will hit the roof, and not the windows, keeping your home cool in the summer.

Solar thermal energy

Solar thermal energy is sort of a combination of photovoltaic and passive solar energy: It uses panels that have water or *glycol* (an antifreeze-type liquid) running through them to capture the heat from the sun. You can use this heat for different purposes, depending on the solar thermal system.

The most common use for solar thermal energy is to heat things such as swimming pools (by using low-temperature collectors) and residential hot water tanks (by using medium-temperature collectors). Medium-temperature solar thermal energy collectors can heat more than water, however; they can even heat large commercial or industrial buildings. Take a look at Figure 13-3 to see how solar thermal energy works.

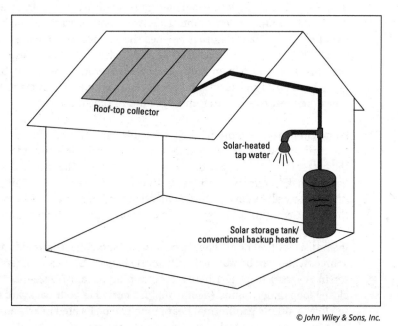

Roof-top collector

Solar-heated
tap water

Solar storage tank/
conventional backup heater

FIGURE 13-3:
Solar thermal
energy in action.

You need other means of energy to heat water — solar heat can't do it all. Depending on how sunny it is and how much hot water you use, solar hot water heaters can supply, on average, 60 percent of the heat you need for your hot water, according to the U.S. Department of Energy's National Renewable Energy Laboratory. Surprisingly, 50 percent of homes in the United States and 67 percent of commercial properties are in locations with sufficient sunlight to use solar thermal energy.

China, Taiwan, Europe, and Japan use the majority of solar thermal energy, and they use it for hot water and space heating. In Canada and the United States, the technology is mostly used to heat swimming pools, though more homeowners are getting on board (especially when they find out they can pay the system off in five or six years if they have a family of four or more).

You can also use solar thermal energy to produce electricity. High-temperature collectors, which take the form of multiple mirrors, concentrate sunlight and focus that energy on a small area. The resulting heat can produce steam from water, which you can use to generate power.

Heat from the ground up

You know the Earth's surface is heating up — that's why you're reading this book. But beneath the round, deep down, it's already hot. Geothermal power derives from the heat beneath the Earth's crust. (Geo literally means Earth, and thermal means heat.)

People can harness geothermal energy through water (or an antifreeze-type liquid) that they pipe underground. That water boils because of the heat from magma (what volcano lava is before it reaches the surface) deep inside the Earth. Magma is incredibly hot — around 1,800 degrees Fahrenheit (1,000 degrees Celsius). The pressure from the steam from the boiled water drives itself through installed pipes back up to the surface and propels a turbine, which creates electricity. The pipes then return the water underground to repeat the cycle.

Some hot spots are more readily accessible than others. The closer a hot spot is to the surface, the easier it is to access. Also, the easier it is to get through the ground, the better chance you have of accessing a hot spot. The United States has the most geothermal resources, with Latin America, Indonesia, the Philippines, and East Africa also well endowed. New Zealand gets 17 percent of its electricity from geothermal. Geothermal energy plants exist on every continent in the world.

Even in areas that don't have access to hot spots, geothermal is a growing energy source. Water can be warmed sufficiently using a heat exchanger (like a refrigerator in reverse) to heat a house by pumping an antifreeze-like liquid through a closed loop underground (a loop of pipes buried in your backyard or under a parking lot). A simple geothermal heat pump can be located nearly anywhere on earth. This form of geothermal energy doesn't make electricity, but it can replace a fossil fuel furnace.

About 65 percent of Iceland's electricity comes from geothermal energy. The country has 200 volcanoes and 600 underground hot springs that it can tap into, enough to fuel all of Iceland's electricity. In fact, the country now has five major geothermal plants. The water itself heats about 90 percent of homes in Iceland and provides all the hot water. Geothermal utilities use the same source to create the many bathing pool hot springs in the country. Iceland has been shifting its electricity technology from oil to geothermal ever since the oil crisis in the 1970s; the country has invested about $8 billion over the past three decades, and it has become almost entirely self-sufficient. The Philippines are close behind Iceland, generating more than 27 percent of their electricity from geothermal energy.

The IEA estimates that the world offers a potential of 85 gigawatts of geothermal energy (a *gigawatt* is 1 million watts), which is 0.6 of a percent of humanity's current energy demand. Geothermal energy sources currently supply the world with more energy than solar and wind energy sources combined.

Geothermal power does have its downsides. Installing a system takes a long time, and the drilling is costly — similar to drilling for oil, but without as big a financial payback. Additionally, concern exists that drilling for geothermal energy can cause earthquakes because every geothermal hot spot is in a geologically active area. Supporters of geothermal energy contend that these quakes would likely be so small that you wouldn't be able to feel them. Others maintain that not enough evidence exists to say that geothermal drilling causes earthquakes at all.

Another way to take advantage of the Earth's underground warmth is through *earth energy*, which doesn't take its heat from magma, but from natural underground hot springs. Pumps can tap into these waters and use the warmth to heat water or the interior of a building. We discuss these heat pumps in Chapter 14.

Hydropower

Anyone who's been to Niagara Falls can attest to the beauty and majesty of rushing water. It's also a great way to generate energy. *Hydropower* uses the flow of water to turn turbines, which convert the energy into electricity. This method is very similar to how a wind turbine generates electricity.

People can generate hydropower in two ways:

>> **Impoundment systems:** Water is stored in dams and reservoirs, which hydropower utilities can then release to help meet power demand at specific times.

   Unfortunately, building large dams can flood natural ecosystems and even nearby communities. The IEA also reports such problems as increased fish deaths and land erosion. (For an example of a problematic impoundment system, see the nearby sidebar in this chapter.)

>> **Run-of-river hydropower plants:** Water's natural flow is used to produce power continuously. The huge power plants at Niagara Falls, for example, are run-of-river. If you have a stream or river on your property, you might be able to use it to provide some or all of your own house's electricity needs. The run-of-river hydropower systems cause very little environmental damage.

Hydropower plants can play a huge part in reducing fossil fuel use. Nearly 70 percent of Brazil's electricity comes from hydropower. Consequently, the country's power sector produces four times less carbon dioxide than comparable power sectors in other countries. On another positive note, hydropower *from existing plants* is one of the cheapest renewable energy options available today because developers and power providers have already developed and built so much of its technology and infrastructure, which are in wide use.

DAM IT

Looking over the Yangtze River, the third longest river in the world, Chinese developers (envisioning a huge source of hydropower) said, "Dam it." The Three Gorges Dam on the Yangtze River, a 15-year, $25 billion project completed in 2012 is the world's biggest hydropower dam. The dam's impact on locals and on the environment has been widely criticized. It remains controversial having displaced 1.3 million people, while being blamed for increased earthquakes.

This example shows that even renewable energy sources can have their social and environmental issues. Future hydropower developers need to closely consider the whole picture before they implement large structural changes.

New big hydro is very expensive indeed, takes decades to build, and is likely to be considered only where no other sources of energy are available.

Ocean power

If you've ever spent time on the sea coast, you know that the tides come in and flow out in a daily cycle, pulled by the moon's gravitational force. You may find yourself scooting your towel up the beach a few feet while the hours of the afternoon go by. Not only can the tides move you off the beach; they can move turbines, too.

Ocean power functions in basically the same way as hydropower, using the force created by the movement of water. But rather than coming from river flow, this water power comes from the movement of the currents, tides, and waves. Here's how each works:

- >> **Currents:** Turbines are placed in flow regions of naturally occurring strong currents.
- >> **Tides:** At full tide, water is held back with gates. When the ocean reaches low tide, the gates are lifted and the water flows out forcefully, spinning turbines to generate electricity.
- >> **Waves:** Turbines are put in the areas of strong wave action, and each wave that hits the turbines spins them.

As with almost all renewable energy, tidal power sources have a high start-up cost, but the environmental benefits could be huge; tidal and ocean energy give off no emissions. For countries that have long coastlines, including Canada, the

United States, and Australia, ocean power holds huge potential. France and China are already using tidal power.

Tidal technologies are still being perfected, however. Ocean power developers are currently working on new pilot projects that use more efficient technology that does not involve damming bays or estuaries.

From plants to energy

Any herbivore or vegetarian can tell you that plants are full of energy, but some plants can power more than just people and animals. When you ferment plants that are high in sugar, such as corn and sugarcane, they form a kind of alcohol, known as *ethanol*, that you can use as a fuel (or *biofuel*, meaning it comes from living organisms).

REMEMBER

Hypothetically, scientists consider biofuels to be zero emission. Even though engines emit GHG when they burn the fuel, that process works in a closed-loop cycle: The carbon dioxide going into the atmosphere is the same carbon dioxide that the plant absorbed from the atmosphere when it was growing. Biofuels also release less carbon dioxide when burned than conventional gasoline. Those biofuels that aren't truly zero-emission are those that use a lot of fossil fuels in the growing process. That's why corn ethanol is less efficient than sugarcane.

Consternation over corn

Ethanol from corn is controversial because it uses a global staple food for fuel. Corn-based ethanol distorts the market because some farmers plant less of one crop (wheat, for example) to benefit from government subsidies for growing corn or grain for fuel. These subsidies are a real factor in the United States and Canada. (We talk about government subsidies in Chapter 10.)

Almost everything people eat nowadays connects to corn somehow. You can break corn down into many forms, including corn flour, corn oil, and corn syrup. You can find it far beyond breakfast cereals — you munch on corn when you eat licorice, table syrup, ketchup, and beer. Behind the meat counter, the beef, pork, chicken, turkey, and even fish that you buy often eat corn feed. Even your chicken nuggets are about 75-percent corn. Forget food — you can find corn products in your toothpaste and lipstick, and even in your drywall, cleaners, and paper products. . . . Need we say more?

The majority of people in the world — those living in developing countries — depend on corn as the basis of their diet. Consequently, energy experts agree that humanity shouldn't make fuels from food crops, especially corn.

The developers behind corn ethanol say that they won't use corn in the future after they find new sources for ethanol. They assert that corn is like a practice run, aiding them to determine which plant sources work most efficiently in developing fuel. The corn era of ethanol is comparable to the black-and-white era of television — with a technological breakthrough just around the corner. Nevertheless, ethanol producers are still using a lot of corn. Incredibly, a full 40 percent of corn grown in the United States is diverted from food to fuel.

Sugarcane as biofuel

Brazil uses sugarcane as the source of its ethanol production. Sugarcane is a much more effective source than corn, particularly because it produces about seven crops before you need to replant it. For this reason, you need to use less energy to produce it than you do to produce corn. Unlike when you grow corn, you don't need tractors that run on gas or pesticides to grow sugarcane. Sugarcane just grows. Another advantage of using sugarcane to produce ethanol is that sugarcane isn't a staple food crop, so using it for ethanol doesn't risk raising food prices.

Brazil's sugarcane farming is controversial, too, because a relatively small proportion of farmers can clear rainforests to produce sugarcane. (We consider this conundrum in Chapter 12.)

Nothing wasted

One person's garbage is another person's alternative energy source. Although civilization produces a great deal of waste, it can turn some of that waste into fuel for energy. Solid waste (such as plant waste and animal waste, which would otherwise go to the compost or landfill), liquid waste (such as used frying oil), and gases (which our landfill sites emit) all offer power possibilities.

Solid waste: Biomass

Another possible solution to the biofuel problem is to forego fresh plants entirely. People can burn the plant and animal waste that currently packs landfills to produce energy — this material is called *biomass*. (Living organic matter burned for fuel is also considered biomass.) Humans have used the simplest form of biomass for thousands of years — wood for fires. People still use wood in fireplaces and woodstoves to this day, and many countries in the developing world depend on wood for both cooking and heating their homes.

Biomass has evolved beyond wood and fire, however. Today, possible biomass sources include solid plant waste from

>> **Farms:** Corn stalks, straw, manure

>> **Forestry and paper industry:** Bark, wood scraps, sawdust, pulp, wood chips

>> **Home:** Kitchen food scraps, yard and garden clippings, sewage sludge

>> **Vineyards:** Grape waste after the grapes are processed and crushed

Burning garbage is different than burning biomass. Toxic chemicals released from burning garbage can be hazardous to human health.

Burning biomass emits carbon dioxide (and some nitrogen oxide and sulfur dioxide) when burned. Nevertheless, burning biomass releases fewer emissions than burning coal does.

REMEMBER

Burning waste is carbon-neutral only if it comes from plant materials because the carbon dioxide released is only what the plants absorbed when growing.

You can burn biomass along with coal. In this scenario, biomass replaces some — but not all — of the coal burned for energy, lowering the emissions produced. This technology is called *co-firing*. Coal plants can begin co-firing right away because you generally burn both coal and biomass by using boilers that heat water to create steam, which then turns turbines to create electricity. Usually, a coal plant can't replace more than 15 percent of its coal with biomass without losing efficiency. Someone would have to develop a new system designed for biomass to make a plant work efficiently using a higher percentage of biomass.

You can even use biomass right in your own home, never mind what the power plant is doing down the road. We talk more about energy-efficient changes you can make in your own home in Chapter 6. You can also turn biomass into biofuel, which we talk about in the following section.

NOT YOUR REGULAR SMOKEHOUSE

In some situations, using biomass as a renewable resource does more harm than good. Two and a half billion people in the world today use biomass — animal poop, farm crop waste, charcoal, and wood — as fuel for daily cooking and heating needs. In many developing countries, people use fires to cook and to heat their small homes. This smoke creates major health problems for the children living in these homes.

International development organizations are currently working to promote solar cooking stoves as a clean alternative, which provides the added benefit of healthier lungs.

Liquid waste

People use renewable sources to create biofuel, and biofuel can replace petroleum-based fuel in gas and diesel engines. The most common and developed types of biofuel are ethanol (technically called bioethanol) and biodiesel. The section "From plants to energy," earlier in this chapter, discusses biofuel and ethanol in detail; in the following sections, we discuss how you can use waste products for these fuels.

ETHANOL

Ethanol is alcohol based and can be derived from many different kinds of plant material — even plant waste. The plant only needs to contain sugars. This group includes corn, wheat, rice, sugarcane, sugar beets, yard clippings, and potato skins.

You can use straw and switch grass, wood chips, corn husks, and poplar trees as biofuel fodder, too. Because these plants aren't high in sugar, they must undergo a special process involving an enzyme that digests their cellulose and turns it into a sugar. This kind of ethanol is called *cellulosic*. Companies such as Iogen and Shell are now making it commercially available.

Cellulosic ethanol does carry a higher cost than corn and grain ethanol due to an additional stage required in processing, but it has considerable benefits. It makes use of agricultural waste, instead of using the crop itself (which can constrain food supplies). And because it comes from existing waste or a naturally growing source, like switch grass, rather than a cultivated agrobusiness crop, it requires far less energy to produce, resulting in fewer emissions.

REMEMBER

Nothing in nature is really waste, so people need to consider the nutrient value of corn husks and straw going back to the soil versus using it as a biomass for fuel.

BIODIESEL

Biodiesel is oil based, and people can make it from sources such as used frying oil. Aside from deriving energy from a waste product, you also get the extra benefit of smelling french fries from the tailpipe of any passing car.

Gas from garbage

Organic materials are composed largely of *hydrocarbons*, which are made up of hydrogen, oxygen, and carbon atoms. When organic material decays, it releases gases, mainly carbon dioxide (carbon and oxygen) and methane (carbon and hydrogen), made from these atoms. Although it's a potent GHG (see Chapter 2), people such as farmers outside of Ottawa, Canada, are using methane as an energy source, processing their manure and food waste to power a generator; the waste becomes neutral (and a high-grade fertilizer) and doesn't poison the water supply. You can also capture methane from landfills, major composting facilities, or sewage treatment plants, and then burn it to produce energy. Methane does release carbon dioxide, but at a very low level, when you burn it.

GOOD NEWS

Capturing methane gas from a landfill in Idaho already powers 24,000 local homes, and the program is growing. The U.S. Environmental Protection Agency's Landfill Methane Outreach Program supports this project and hundreds of others.

Exploring Another Nonrenewable Energy Source: Nuclear Power

Nuclear technology produces electricity around the world, although it's a hot topic of debate. Now that people recognize the climate change in the air, nuclear power has regained favor among some as a low-emission energy source. But, like big hydro, nuclear may not be a competitive source of power in the future, except in countries that already have a large installed base of nuclear generation. The IEA says the following::

> "The political, economic, social and technical feasibility of solar energy, wind energy and electricity storage technologies has improved dramatically over the past few years, while that of nuclear energy and carbon dioxide capture and storage (CCS) in the electricity sector have not shown similar improvements."

Desperate to find ways to get off fossil fuels, the nuclear industry hopes it may get a new lease on life. With Germany closing reactors and no new reactors having been built in North America or Europe for more than a decade, nuclear wants to be a climate solution. The next sections examine the pluses and minuses.

Understanding nuclear power

Nuclear reactors produce electricity through *nuclear fission* (or splitting the atom). Current conventional reactors use uranium, a naturally occurring radioactive mineral, as the fuel for the chain reaction of nuclear fission, when one molecule blasts off an electron to split another molecule. The heat generated by the chain reaction boils water to create steam, which turns turbines to make electricity. Essentially, nuclear power is just a tea kettle on a very dangerous nuclear fire.

New technology, as yet noncommercial and unproven, is also being touted as a replacement for fossil fuels. Generally under an umbrella term of *small modular reactors (SMR)*, the industry is promoting a range of designs that exist experimentally or only on paper. Canada's government with the province of New Brunswick is moving ahead with a British manufacturer to build a molten salt reactor near the existing CANDU reactor at Point Lepreau.

Nuclear power is a nonrenewable resource because the Earth has a finite supply of uranium, although it has more uranium than fossil fuels. Uranium has its definite drawbacks as a material, which we discuss in the section, "Weighing the negatives," later in this chapter.

Looking at the positives

Nuclear energy's supporters cite the following benefits:

>> **Low GHG emissions:** Nuclear power plants produce only indirect emissions, relating to mining and transporting the uranium, building the plant, and (depending on the type of reactor) enriching the uranium.

>> **Mature technology:** Unlike the other energy sources we discuss in this chapter, the infrastructure and systems to support nuclear power plants and related mining activities are already in place in some countries.

>> **Steady prices:** The price of nuclear energy doesn't fluctuate as much as the price of energy generated from fossil fuels. (The cost of constructing nuclear generators, however, continues to rise.)

Weighing the negatives

Detractors of nuclear power voice major concerns, including the following:

>> **Health concerns:** Long-term health studies vary, but a number of recent studies demonstrate higher cancer rates among populations that live near nuclear reactors. Uranium mining also endangers the health of miners.

>> **Reliability:** Nations around the world have varying success with reactor reliability. Some countries, such as Canada, have had persistent problems keeping reactors on line. Breakdowns and retrofits have cost taxpayers billions of dollars.

- >> **Risk of proliferation of nuclear weapons:** People can use nuclear fuels in nuclear weapons. India made its first nuclear weapon by using spent fuel from a Canadian reactor.

- >> **Safe storage of nuclear waste:** Nuclear waste must be kept out of the biosphere for at least a quarter of a billion years before it's no longer toxic. No current technology can contain nuclear waste for that length of time.

- >> **Safety of nuclear plants themselves:** Many people equate nuclear power with the Chernobyl disaster, the near-meltdown of a nuclear power plant in the Soviet Union in 1986. Fears exist that a major accident could happen again. Nuclear reactors do run an extremely small risk of experiencing a catastrophic accident.

- >> **Security concerns:** Some people worry that terrorists could target nuclear plants and materials because of the great and long-lasting damage such an attack could cause.

But the main reason that nuclear power is unlikely to be a big part of the energy revolution is simply because it's far more expensive than most alternatives and takes a very long time to build out. Existing nuclear plants will keep working until the end of their design life, but new installs will be few and far between.

The member countries to the Kyoto Protocol officially decided that nuclear power isn't clean enough for the CDM. The European Union voiced a similar view of nuclear energy. Working toward a goal of getting 20 percent of energy from renewable sources by 2020, Germany, Spain, and Sweden are committed to shutting down their nuclear plants. France relies on nuclear energy in a big way, however, and so does Switzerland.

Chapter **14**

Show Me the Money: Business and Industrial Solutions

C limate change is a primary strategic issue for businesses around the world. Fossil fuels will no longer be the world's primary energy source, which will affect enterprises at all levels and in all economic systems.

The effects of climate change on different businesses will vary a great deal, but almost all energy suppliers and energy users will be making dramatic changes. One thing you may hear some businesses tell you is that they can't afford to reduce greenhouse gases (GHGs) and switch to sustainable energy. If people expect companies to spend a fortune on reducing GHGs, business reps say, those businesses will be hobbled in today's competitive marketplace. Other businesses are grabbing the chance to make money by making their processes more efficient or making their products less carbon-intensive. Those in the fossil fuel industries can see the writing on the wall — their primary reason for doing business is disappearing fast, and their whole enterprises are in danger. Power utilities both make power and distribute it — they have opportunities in the coming world where the sources of electrical power will be widespread, but their big central plants may not be competitive when cheaper renewables come to dominate the market.

This chapter describes some of the actions that have been taken and are being taken now, and speculates a bit about what businesses might do in the future.

In industrial economies, companies can improve how they manufacture products, using modern energy-efficient equipment and recycling. But more than just manufacturers and oil companies can get in on the game. Companies can also get involved in the creation of new green services or in the carbon market. They can change how they construct buildings or turn wood into paper. And you can help them by demanding new, greener products that don't produce as much GHG, and by rewarding those companies that put sound GHG–fighting practices in place. You're their customer — and the customer is always right.

In developing economies, the promise of abundant low-cost renewable energy is that they may be able to leapfrog many of the technologies embedded in the industrialized world and go immediately to newer more energy efficient technologies. New business opportunities can develop rapidly when there is no entrenched infrastructure. More on this in Chapter 17.

Processing and Manufacturing Efficiently

Most manufacturing requires a great deal of energy, usually from fossil fuels that create a lot of carbon dioxide emissions. In many cases, much of this energy is actually wasted, thanks to old and inefficient equipment and weak regulations governing its use. The actual manufacturing process creates even more GHGs.

REMEMBER

Energy intensity measures the energy required to produce economic output. It combines measures about conservation, or *wasting* less, with measures of efficiency — *using* less per unit of production. Reducing energy intensity is just good business practice, with or without climate change. So energy intensity has been dropping for years, as enterprises respond to commercial needs. Now, reducing emissions is just as vital. Businesses around the world are making money by combining reduced energy intensity with reduced emissions.

Taking steps to conserve energy

Manufacturing doesn't have to be wasteful; with some tweaks, industry can use less power, causing fewer GHG emissions. There are numerous environmental consulting firms all over the world, as well as nongovernmental organizations (NGOs), that specialize in working with businesses, companies, and industry to reduce their GHG emissions by lowering energy use.

Here are a few steps that the Intergovernmental Panel on Climate Change (IPCC) recommends manufacturers take to help conserve energy:

>> **Measure how much energy the manufacturing process uses and how many emissions it creates.** Use this information to set benchmarks and goals for reduction. The industry can measure its success from the changes it makes.

>> **Use the correct-size equipment properly and conservatively.** Keep the equipment tuned up and fix malfunctions when they occur. Also, choose equipment such as the optimal size of piping to cut energy use.

  You'd be surprised how often mismatched pipes and oversized motors waste large amounts of energy. Energy guru Amory Lovins, of the Rocky Mountain Institute, estimates that the United States could cut electricity use by 40 percent simply by replacing the wrong-sized electrical motors with motors that are the correct size!

>> **Use more energy-efficient motors and equipment.** Running motors accounts for about 45 percent of electricity use in European, Canadian, and U.S. industries. Businesses can run motors more effectively by changing materials and improving aerodynamics. Businesses can improve the efficiency of all sorts of equipment — even fans and pumps — along the way. Industry sources suggest that newer motors could reduce worldwide electricity consumption by 10 percent.

>> **Insulate buildings and equipment sufficiently. Insulation keeps any building's energy use low, whether you're trying to keep the heat in or out.** Likewise, insulating hot water pipes reduces the loss of heat and also lessens the energy needed to heat the water and compensate for the lost heat.

>> **Reduce leaks of any sort (such as air and steam).** For example, when the pressure of the steam drives a turbine, escaping steam reduces efficiency. The boiler has to run that much harder to make up for lost steam. Air leaking into boilers and furnaces can have the same impact. Leaks mean that energy is being spent driving the air or steam into places it shouldn't be.

>> **Recycle materials.** Both the steel and aluminum industries have found recycling to be a major advantage. Recycling the steel from old furnaces, for example, makes up a whole third of global steel production and uses 30 to 40 percent of what the process takes if started from raw materials.

Not only can these steps help companies cut back on their GHG emissions through lower energy use; they can also save companies money by no longer wasting pricey power.

Using energy efficiently

With the money saved through energy conservation, companies can adopt new, efficient technologies for applications such as electric equipment, heaters, and boiler systems:

>> **Systems powered by sustainable energy:** Industries can use their own biomass waste, such as wood, food, pulp, and paper scraps, as fuel. Some industries can power themselves by using methane from landfills to run boilers. Solar and wind power are other renewable options. (Refer to Chapter 13 for more about sustainable energy.)

Not every industry can turn to sustainable energy; some manufacturers require a particular fuel, such as the steel industry, which uses coking coal. Mini-mills now make steel from recycled metal using electric arc furnaces, a sign of things to come. (Refer to Chapter 5 for more about the steel industry.)

>> **Combining heat and power:** Businesses can use up to 90 percent of the excess heat given off by power production (or generated by machinery) to replace regular heating within a building or buildings, instead of simply pumping that heat out of the building. Businesses in Germany and the Netherlands use this technology, known as *co-generation.*

High-efficiency, new technology can save companies money in the long run, reducing their energy consumption considerably, but the technology isn't cheap to obtain. Companies in developing countries, especially, might balk at the expense of high-efficiency equipment, and understandably so — developing countries often simply don't have the budgets and financial support for energy-effective technology. At a conference in Copenhagen in 2009 the rich countries of the United Nations committed $100 billion per year in assistance to developing countries to help with this. So far, that commitment, which was to start in 2020, hasn't been matched by action — the start date has been shifted to 2023, and very little money has actually been transferred.

Subsidies from national governments for energy-efficient practices can help industries make big changes in how they operate. Programs following from the Kyoto Protocol and the Paris Agreement are intended to help developing countries obtain these two new technologies:

>> **The clean development mechanism (CDM):** Under this program, industrialized countries pay for clean energy projects in developing countries.

>> **Joint implementation:** Through this program, industrialized countries and developing economies partner to implement projects such as capturing methane from landfills and using it to produce energy, or shifting from coal to renewable energy sources.

Both programs help industries in the developing world introduce efficient technologies and reduce greenhouse gas emissions. (We cover the Kyoto Protocol and the Paris Agreement in Chapter 11 and discuss their relevance to developing countries in Chapter 12.)

Considering individual industries

Although the steps laid out by the IPCC (which we discuss in the section, "Taking steps to conserve energy," earlier in this chapter) are relevant to most manufacturers, specific industries face particular challenges. Chapter 13 discusses renewable energy.

Cement production, for instance, is a particularly carbon-intensive industry, making up a whopping 8 percent of all carbon dioxide emissions in the world. Parts of these processes are unavoidable, such as the carbon dioxide that the limestone of the cement naturally gives off when the cement forms. However, the industry could benefit from using energy from renewable resources or from systems that capture and store the carbon dioxide emissions underground. Companies and institutions around the world are researching ways to make concrete without such massive emissions. The work has been moving rapidly with excellent results — the world market for green concrete is projected to reach $25 billion by 2024.

Other GHG-intensive industries have found some innovative ways to reduce their emissions. Here are a couple:

>> **Pulp and paper:** Canada's forestry giant Tembec has managed to reduce its production of GHGs directly and indirectly. It recycles wood chips and other waste as fuel, burning them in place of higher carbon-emitting fossil fuels. (This kind of fuel use is sometimes called a *closed loop system,* in which you don't

input to or output from the system — it's a full cycle.) Tembec now dries its pulp by using more efficient electric hot air dryers, replacing earlier steam units that were fossil-fuel powered. Tembec even extracts sugar from the *cooking liquor,* the fluid left over after paper is manufactured, and turns it into ethanol, which it then sells as a product to be used in things like antiseptics and sanitizers.

GOOD
NEWS

Thanks to steps such as the ones Tembec has taken, the Forest Products Association of Canada boasts that it has met and far exceeded the Kyoto target of reducing carbon dioxide emissions to 6 percent below 1990 levels. Canadian pulp and paper mills have actually reduced their emissions by 66 percent below 1990 levels.

>> **Aluminum:** Many aluminum companies are concentrating today on recycling. To recycle aluminum, these companies need to use only 5 percent of the energy they would need to manufacture it from raw ore. One manufacturer, Alcoa, plans to boost the percentage of recycled aluminum it uses in new production to 50 percent by 2050, and to reduce GHG emission intensity by 30 percent by 2025 and 50 percent by 2030 from a 2015 baseline.

GOOD
NEWS

Thanks to the introduction of newer technology, the aluminum industry has been able to reduce its production of perfluorocarbons, the particularly nasty GHG that captures from 6,500 to 9,200 times more heat than carbon dioxide.

Trading Carbon between Manufacturers — The Carbon Market

Some manufacturers and producers can reduce their GHG emissions more easily than others. To address that imbalance and enable industries to reduce their GHG emissions across the board, some jurisdictions and commodity traders have created carbon markets. The *carbon market* isn't like a flea market, the sort of place you drop by on a Saturday morning to pick up a bargain in chunks of coal.

Here's how a private carbon market works:

1. **Companies form a group and make a commitment to each other.**

2. **They agree on how they want to reduce their emissions over the year individually and collectively.**

3. **If the company reduces its emissions more than planned, it has *carbon credits,* which it can sell.**

   If a company doesn't make its goal, it can buy someone else's credits. (Another name for this process is *cap and trade.*)

Companies can actually make money by reducing their output, creating carbon credits and selling them. The carbon market ensures that carbon dioxide levels are being reduced. It just doesn't worry about where or by whom.

Although the majority of carbon markets around the world are government initiatives, businesses can and have implemented emissions trading themselves. (We discuss government-led carbon markets in Chapter 10.)

Beyond do-gooding, businesses can make money in carbon trading. We talk more about how banks are getting involved in fighting climate change — and profiting, in the process — in the section "Focusing on Support from the Professional Service Sector," later in this chapter.

ADAPTATION FOR INDUSTRY

Industry is as vulnerable to the effects of climate change as are communities and individuals. Businesses need to adapt to these changes, just like everyone else.

In part, they need to make mundane, obvious adaptations — renovating facilities to anticipate extreme weather, for example.

Mining companies are re-engineering *tailing ponds*, the holding ponds for mine waste, to prepare for more extreme weather events. Diamond mines in the Arctic are adapting to the warming weather — the winter roads, relying on consistently frozen ice and snow, on which they used to rely are now available for less of the year, due to climate change. This is driving up their costs because they have to fly more goods in.

Industry can also make proactive adaptations, however, involving diversifying and thinking in new ways. BP, also known as British Petroleum, is a prime example of smart adaptation. By broadening its mandate beyond oil, it can supply the market's energy needs, whether the world depends on oil, ethanol, solar power, or wind energy.

Any major production or manufacturing company can take similar actions. Forestry and agriculture-based companies, for example, can diversify their resource base so that they harvest from numerous small locations, rather than one large location. This adaptation can allow them to continue supplying wood to the market, even if climate change and its effects impact one area, for example, with forest fires or droughts.

Constructing Greener Buildings

Companies also can make a difference by changing how they do business, but also where they do it — their offices and factories. These big buildings currently emit a lot of GHG, but that means they also present a big opportunity. In fact, the IPCC reports that improving the efficiency of commercial and industrial buildings is a more cost-effective way of reducing GHGs than overhauling industry's manufacturing processes. The following sections describe some of the ways that construction of buildings can affect emissions.

Cutting back on heating and cooling

Businesses often use the most energy keeping an office or factory warm in the winter or cool in the summer. Those businesses can cut back on energy consumption by installing heating and cooling systems that don't guzzle so much (or any) carbon-emitting fuels. In Chapter 6, we look at different heating and cooling options for houses. These solutions also apply to companies. Companies can reduce their heating- and cooling-related energy consumption in some very simple and inexpensive ways:

>> **Insulation:** A properly insulated building doesn't require nearly as much energy to heat or cool. You lose about 60 percent of the heat pumped into a poorly insulated building through the roof and the walls alone.

>> **Air circulation:** Buildings in moderate climates can benefit from using systems that either keep the outside air out or let it in, depending on whether you need to heat or cool the building on that day.

>> **A green roof:** A rooftop that features layers of grass, soil, and waterproof lining naturally cools a building, providing insulation and reflecting the sun's light (unlike black asphalt roofing, which absorbs it). In fact, a green roof reduces air-conditioning demand by 25 percent. The plants on these roofs have the added bonus of absorbing some extra carbon from the atmosphere.

>> **A white roof:** For roofs that can't support plant life, a simple lick of white paint can help cool buildings in hot climates because the white paint reflects light.

Good insulation coupled with good natural lighting can even heat a whole building. The Rocky Mountain Institute in Colorado is (as the name suggests) located in the Rockies and is under heavy snow much of the winter, yet the building has no furnace. It relies instead on the sun's rays, which stream in through giant windows, and it holds that heat through maximum insulation. They even grow bananas indoors to show the building's balmy conditions. Check out founder Amory Lovins's website (www.rmi.org) for more wonderful information on the enormous potential of energy efficiency.

Exploring energy alternatives

Business and industry can lead the charge in shifting buildings away from consuming fossil fuels and toward renewable resources. Because their buildings consume so much energy, when they make a change, it has a big effect. And as the cost of renewable power continues to decline, more and more corporations from Apple and Amazon to Verizon and Walmart are using solar energy generated from their own rooftops to reduce their energy costs, and in the process they're making real contributions to reducing carbon emissions. Airports and shopping malls are doing the same. Google alone purchased 1,600 megawatts of solar power in 2019, more that the output of all but the very largest hydroelectric dams. For an overview of the energy alternatives that businesses can investigate, check out Chapter 13.

REMEMBER

By creating its own energy from renewable sources, a building can remove itself from the electricity grid — or, at least, reduce its use of energy coming from the grid. The more energy a building can produce on its own, the more money the owner saves. In some jurisdictions, building owners can even sell their excess renewable energy back to the grid and make a profit.

Certifying new buildings

Improving efficiency and energy sources can cut back on the GHG emissions of existing business buildings, but when companies require new buildings, they have the opportunity to really go green. They can ensure that their new digs are as GHG-smart as possible by following environmental standards. Having a standard for buildings sets the bar for companies. And while companies improve and compete, they raise the bar by raising standards.

In North America, LEED (Leadership in Energy and Environmental Design) leads the way. The United States Green Building Council created and constantly modifies the LEED to reflect new practices and materials. The LEED provides a set of standards that new or renovated buildings must achieve to be certified. LEED standards encompass all aspects of a building's construction, including the following:

>> The percentage of demolition material recycled (in the instances when a building was demolished to make way for new construction)

>> How efficiently the building uses water

>> The materials used and what percentage of those materials came from less than 500 miles away

LEED also has a certification program for architects and engineers to confirm their understanding of green building products and practices. To find out more about LEED in the United States, see www.usgbc.org. For LEED in Canada, check out www.cagbc.org.

Other countries have programs similar to LEED:

>> **United Kingdom:** The Building Research Establishment and Environmental Assessment Method (BREEAM) is the first set of environmental building standards established anywhere — developed even before LEED. BREEAM has specific requirements for the widest range of buildings — from schools to warehouses to theaters. They even have a ranking for prisons! (You can visit the BREEAM site at www.breeam.org.)

>> **Australia:** Green Star is run by the Green Building Council of Australia. Green Star was inspired by BREEAM and LEED, but modified to better fit the Australian environment. Its particular focus is office buildings, from design, to construction, to retrofits. Green Star hands out stars (up to six) to buildings that meet its environmental standards. It plans to expand its standards to cover industrial, retail, and residential buildings. (For more info, visit www.gbca.org.au.)

GOOD
NEWS

Other companies and architects all over Earth are designing and constructing new efficient green buildings. Check out www.re-thinkingthefuture.com/rtf-fresh-perspectives/a1057-20-highest-rated-green-buildings-in-the-world/ for a summary.

Identifying Corporate Success Stories

If you zone out whenever you hear the word "corporate" because it seems so big and heartless (or just plain boring), we want to change that. Corporations are using their big profiles (and big budgets) to help reduce emissions. Table 14-1 profiles some of the corporations that are ahead of the pack in reducing their carbon footprints.

TABLE 14-1   **Businesses Reducing GHG Emissions**

Company	Target	Successes or ?
Barclays (banking)	Net zero by 2050	Climate dashboard on public website and £100 billion green financing commitment by 2030
DuPont (chemical manufacturer)	Reduce GHG emissions 30 percent including sourcing 60 percent of electricity from renewable energy, and deliver carbon neutral operations by 2050 or sooner	–67 percent GHG emissions, +35 percent in production, and +8 percent in renewable energy
HDR, Inc. (architecture and engineering)	Use sustainable development principles in all projects	11 percent emissions reduction 2011 to 2018, 5.8 million sq ft of LEED buildings
Intel (computer manufacturer)	–10 percent PFC* emissions by 2010 (versus 1995) and –4 percent energy use by 2010 (versus 2002)	Emissions actually rose from 2018 to 2020
Interface (floor-covering manufacturer)	Zero environmental impact by 2020	–45 percent energy use, –60 percent GHG emissions (versus 1996), and +16 percent use of renewable energy
Johnson & Johnson (health care products)	–7 percent GHG emissions by 2010 (versus 1990), commitment to 100 percent renewable electricity by 2023	Carbon dioxide emissions down from 1.2 to 950 million tonnes, 54 percent of electricity from renewable energy
Toyota (car manufacturer)	To become an environmental leader	–2.5 percent carbon dioxide emissions in 2 years and –35 percent in energy use in 5 years, $180 million fine for noncompliance with emissions regulations
Wal-Mart (discount store)	–30 percent energy use, –20 percent GHG emissions in 7 years, and use of 100 percent renewable energy by 2035	$500 million annually invested in energy-efficient technologies, procured most wind energy in the United States

Recognizing Corporate Nonsuccess Stories

The success stories in the preceding section all are from energy *consumers*.

Energy *producers* face different challenges. Some companies that are deeply embedded in fossil fuels have made attempts to change their strategic direction, but have found it difficult. See the nearby sidebar for an example.

BP'S ATTEMPT TO GO "BEYOND PETROLEUM"

The best known example of this is British Petroleum (BP). John Browne was CEO of BP for 10 years from 1997 to 2007 — he was wise enough to see that his company wasn't acting responsibly in encouraging more production and use of carbon-based fuels. He tried to change the company's direction, even rebranding the company as "Beyond Petroleum." The rebranded BP made strategic investments in renewables and carbon capture technologies.

Unfortunately, its competitors didn't care as much about being socially responsible — Exxon in particular continued to lead the industry pushback against well-known climate science and continue to invest in unregulated fossil fuel activities. And Exxon made more money for its shareholders. BP had to climb down from its socially responsible high horse and has carried on a more traditional oil and gas company (although all these companies make pretty loud noises about a transition, there's little evidence that it is actually happening).

Likewise, big energy utilities are having real difficulties adjusting their businesses to the new reality. Imagine, for instance, you're on the board of directors of British Columbia Hydro. You're spending $14 billion on a huge new dam and generator project to make about 1,400 megawatts of electrical power. Your engineers project that you'll want to sell the resulting electricity for around 14 cents a kilowatt-hour to break even. And you see from the news that electricity from solar PV is now selling for less than 3 cents a kilowatt-hour in Arizona and 4 cents in neighboring Alberta. And you're connected to the same grid. How are you going to compete? What are you going to tell your lenders?

In these and many other cases, the existing organizations depend on outdated business models — the fossil fuel producers (just like asbestos companies, leaded gas producers, and tobacco companies) depend on being able to sell products that are dangerous to the health of people and of the biosphere. They know their days are numbered, but they'll fight as hard as they can so long as they can still make a profit.

The business model for energy utilities comes from the old days when it took millions (now billions) of dollars to build big energy projects (using coal or hydro or nuclear), and more millions (now billions) to build the network to distribute the power to consumers. So they had to borrow all that money up-front to build the projects and the networks. To make the borrowing possible, governments gave utilities monopolies — they would be the only organization allowed to sell electricity in a defined area, so they could be guaranteed a profit. That enabled the utilities to borrow money to finance their projects.

Imagine yourself sitting in that same boardroom now, contemplating a world where your retail customers have solar panels on their roofs and want to sell *you* power, where other producers of cheap renewable energy want access to your expensive distribution grids, where your previous generating facilities now are best used as colossal batteries, and where your product is rapidly becoming too expensive to sell. Perhaps you wouldn't want to be in the business anymore? But how can you get out? You still owe your lenders all those billions. You can't make your primary product any cheaper. Oops. Do you or your bank or pension funds have investments in these utilities? Refer to Chapter 17 for more information about this dilemma.

Focusing on Support from the Professional Service Sector

Although some people might say that bankers, insurers, and lawyers produce a lot of hot air, their businesses don't immediately come to mind when you think of global warming. Nevertheless, companies in those sectors have seen the silver lining of profitability in the climate change cloud and have gotten involved with ventures that help others cut back on carbon. This section describes some of these initiatives.

Banking on the environment

Many banks are becoming involved with the fight against climate change by offering specialized services to clients committed to reducing GHGs or providing renewable resources. American investment bank Goldman Sachs says that by 2021, investments in renewable energy overtook the amount being invested in fossil fuels. Further investments in renewables will amount to over $16 trillion before 1930. Here are some examples:

>> **Goldman Sachs:** Goldman Sachs states that the world is at an *inflection point* (a point of a significant change in the shape of a curve) for the deployment of renewables. The company has invested heavily in carbon markets, owning large shares in the European Climate Exchange. (We talk about private carbon trading in the section, "Trading Carbon between Manufacturers," earlier in this chapter.) Goldman Sachs also founded the Center for Environmental Markets, which issues grants for research in market solutions for environmental issues. The company is targeting $150 billion for investments in renewable energy sources.

>> **Bank of America:** This U.S. bank has created an environmental banking group dedicated to conservation and reducing global warming. The bank has recently announced a goal of deploying $1 trillion in its Environmental Business Initiative.

>> **ABN AMRO:** This Dutch banking giant offers risk management services to help its clients reduce their possible losses from climate change. It has made a commitment of 425 million euros ($480 million) to its Sustainable Impact Fund.

Stranding assets and liabilities

Fossil fuel companies are being disrupted by changes in policy, shifts in demand, and new hard-to-beat competition by lower-cost low-carbon alternatives. Banks have billions of dollars in credit extended to those companies. That credit was based on the security of the fossil fuel assets that the companies owned or licensed. Mark Carney, when he was Governor of the Bank of England, warned that those assets would become almost without value, "stranded" when the realities of climate change were recognized. Carney and others estimate that the total amount that may eventually disappear from fossil fuel companies' balance sheets could be from $2.5 to $25 trillion worldwide. That could lead to massive upheavals in world financial markets.

The banks who have extended money to these companies are in a quandary: Do they give more money to those customers in the hope that they'll be able to make enough profit to pay the loans back before their assets lose value, or do they cut them off now and write off the loans? Do you have pension fund or 401(K) in a similar position?

Bankers and financial regulators around the world are requiring companies to evaluate their business prospects to reflect the risk that some of their assets may be worth less because climate risk may cause them to be "left in the ground." Companies are also changing their internal practices, with or without pressure from regulators. In 2020, Royal Dutch Shell wrote off $22 billion in assets that the company could no longer count as valuable. BP likewise wrote off $17.5 billion.

That's giving investors vital information to judge their own exposure. Consider this: From 2016 to 2020, 27 of the world's 60 largest banks decreased their lending to fossil companies. That's in addition to the $14 trillion of divestment (*divestment* is the act of selling off investments — in this book, it's specifically about reducing investments in fossil fuels assets). Refer to Chapter 6 for more discussion.

These trends are accelerating. Banks that look to the future will do better than those that stay mired in the past. But the pain of the change is real. That's why you should expect continued opposition from the organizations stuck in these old businesses.

Insuring against climate change

The insurance industry has a vested interest in stopping climate change; the extreme weather it brings (which we discuss in Chapter 7) is resulting in a huge surge in claims. Since the 1980s, payouts have doubled every 5 to 10 years. The following are examples of insurance companies investing today to prevent giant payouts in the future:

>> **American International Group (AIG):** The U.S. company offers financial support to projects that encourage GHG emission reductions and now has a specialized division "AIG Energy Renewables" to handle both insurance and investments. AIG says it will invest in forests, renewable energy resources, GHG-mitigating technologies, and green real estate.

>> **Swiss RE:** The giant Zurich-based Swiss insurance company is also interested in working to reduce and profit from the danger of climate change. Swiss Re has committed to getting the company itself to net zero by 2030 and getting its entire investment portfolio to net zero emissions by 2050.

The insurance industry is also actively funding research efforts in these ways:

>> In Canada, insurance companies fund the Centre for Catastrophic Loss Reduction at the University of Western Ontario.

>> In Bonn, Germany, the Munich Climate Insurance Initiative (MCII) is helping to develop alternative insurance products that can facilitate both spreading climate-related risks (ensuring that no one insurance company shoulders the burden of paying out for the aftermath of extreme weather events) and adaptation-response measures. Members of the MCII include the International Institute for Applied Systems Analysis, German Watch, the Potsdam Institute for Climate Impact Research, and individuals from the World Bank and Munich Reinsurance Company. The insurance industry largely funds the effort.

However, on the other hand (there's an "on the other hand" to all these business stories), insurance companies are, after all, in the business of making money. They prefer in all cases *not* to pay claims. So they're now simply refusing to make insurance available to homeowners or businesses in low–lying areas subject to sea level rise or river flooding, or to areas close to potential forest fires and wildfires. More examples are bound to happen as more risks become evident.

REMEMBER

If you own your home, and especially if you're in a high-risk area, you may want to review your insurance policy to ensure you're covered.

Making it legal

Many law firms are actively engaged in making the fight against global warming a legal imperative. Firms such as the U.S.-based Baker & McKenzie offer to help government clients worldwide develop climate change laws and regulations. Firms also offer services on carbon markets, carbon-offset projects, and trading emissions. The legal framework for such projects, even those outside government regulations, requires careful drafting.

GOOD
NEWS

But beyond the services they offer, law firms are businesses like any other. They have an impact through the energy they use and the paper waste that they produce when they work with clients on important issues. The U.S. Environmental Protection Agency and the American Bar Association (ABA) have partnered to address climate change issues through a voluntary program called Green Energy Partners, with more than 1,700 firms, nongovernment and government entities, including more that 40 leading law firms have signed on to one of three commitments:

>> Buying renewable energy

>> Lowering the amount of waste produced

>> Cutting office energy use by at least 10 percent

All the actions in this simple and effective program help reduce the law profession's carbon footprint.

Looking At Farming and Forestry

Farming and forestry are uniquely posed to make a difference in the fight to stop global warming. Like all industries, they can cut back on their GHG emissions by improving their energy efficiency and moving to sustainable energy sources. But what makes them truly exceptional is that they can actually increase how much carbon dioxide is absorbed from the natural greenery and soils under their management. Talk about a global warming one-two punch!

How agriculture and forestry use the land has created a third of global GHG emissions — which means that these industries can become a huge part of the solution to climate change. This following describes just a few of the many ways that improvements to farming and forestry will affect climate change.

Supplying biofuels

The waste that forestry and farming produces, such as wood, crop waste, and manure, doesn't have to be wasted (although leaving waste in the forest or the fields isn't necessarily wasteful — this material can provide nutrition to the land). All that stuff is actually *biomass,* biological material that humans can use, either by burning it to create energy or turning it into *biofuel,* which transportation devices can use. Biofuel gives off fewer GHG emissions than fossil fuels when burned. People are increasingly using biofuel as either an alternative to diesel fuel or as an additive to it, which creates a lower-emission fuel. (You can read up on bio-fuel in Chapter 13.)

Beyond providing the materials for biofuels, agricultural and forestry practices can benefit from using the biofuel themselves. Although companies would have to make an initial minimal investment to convert diesel engines to run on biofuel, this investment would likely offer long-term savings, particularly while oil costs continue to rise.

Improving land management

Land management includes many elements. For forestry, land management involves how and where companies grow the trees and what kind of harvesting methods they use. For agriculture, it involves how farmers till the soil, what they add to the soil, and how they grow and harvest the crops. Land managers can engineer all aspects of their operations to be more environmentally friendly, sucking extra carbon dioxide out of the air.

Forestry

The number-one, land-management solution recommended by the IPCC for forestry is to decrease the areas deforested.

WARNING

The IPCC says that forestry practices must change quickly to counteract rapid, worldwide deforestation. The IPCC warns that forestry companies need to know how climate changes will affect their forests. These changes could include an increase in how fast wood decomposes, as well as more intense droughts and forest fires. (Climate change won't be bad news for all forests initially, however. Because of temperature increases, some trees could grow faster and take in more carbon dioxide.)

Instead of deforesting entire areas, the forestry industry needs to adopt more sustainable methods, such as *selective harvesting*. This method involves removing small groups of trees, leaving behind a range of trees of different ages and sizes. Selective harvesting has many benefits:

>> **Helps land stability:** The root systems of trees hold the soil together and assist in the prevention of landslides.

>> **Keeps the forest functioning as an ecosystem:** This enables the forest to continue to serve as habitat for wildlife.

>> **Keeps the soil healthy and productive:** Healthy soil, supported by trees, takes in rain — a lot of it. Trees also make the soil more drought resistant by shading it and giving it nutrients.

Ecologists encourage selective harvesting in temperate forests primarily to protect wildlife habitat and biodiversity. Temperate forests include the boreal region, a large band of forests, through Russia, Scandinavia, Alaska, and Canada, which makes up about one third of the planet's remaining forests. These forests have deep soils that can support new growth after clear-cutting. The second-growth forest lacks the species diversity of the primary forest, but only very rarely does a clear-cut forest in a temperate region result in true deforestation.

Selective harvesting is essential to the survival of forests in the tropics, however. Tropical forests grow on very thin soils, which are unlikely to be able to support life after the forest canopy has been cut away. Clear-cutting a tropical forest results in true deforestation.

The forestry industry can implement the sustainable practice of lengthening the time between *rotations* (the time between logging the forest, allowing regrowth, and coming back to log it again), allowing a forest to grow for a longer period of time before returning to log that area again. This longer cutting cycle would increase the carbon uptake of the forests; although young forests absorb carbon more quickly than old trees, older trees can retain far more carbon.

GOOD
NEWS

Some in the forestry industry are already taking steps, through the Forest Stewardship Council (FSC), to ensure that forests are sustainably managed. If a forest complies with the FSC's Principles of Responsible Forest Management, the FSC certifies that forest, enabling the operators to use the FSC logo. According to the FSC, more than 200 million hectares have been certified — about 17 percent of the world's industrial forests — and this number is growing rapidly. Large companies dedicated to helping forest management help make these kinds of programs a success — from FSC-certified chairs to eyeliner pencils. (Check out www.fsc.org for more about the Forest Stewardship Council.)

Governments and NGOs are taking steps to reduce deforestation in the tropics, where trees take in carbon dioxide all year round, making deforestation reduction in tropical regions most effective — and most urgently needed. (Boreal trees in the north don't take in carbon dioxide in winter months.) The majority of the world's rainforests are found in developing countries. The Clean Development Mechanism, a program under the Kyoto Protocol, encourages industrialized countries to fund sustainable forestry practices in developing countries that can help those developing countries cut greenhouse gas emissions. (We talk about this program in Chapter 12.)

Farming

Farming may not seem like an obvious culprit when it comes to carbon emissions, but it's a considerable contributor — about 18 percent of global emissions come from farming and land use changes. Humanity's ever-growing food needs and desires have pushed the farming industry to *de*forest valuable land and use emission-heavy methods of farming. The IPCC singles out land use changes, including *re*forestation, as a much needed contributor to keeping to the 1.5 degree limit. Fortunately, greener options are possible.

REGENERATIVE AGRICULTURE

Regenerative agriculture is one of the necessary solutions to the climate emergency. Making agriculture *regenerative* means keeping a vegetative cover on the soil at all times to reduce soil disturbance and leaving more crop residue and waste on the land to return to the soil, to help bring the soil back to health. Modern industrial agriculture uses soil as a structural material to hold plants and applies artificial nutrients and other materials (also known as fertilizer) for plant growth. In the process, soils release carbon to the atmosphere. With regenerative agriculture, soils can return to being a major carbon sink.

In combination with other practices that we discuss in the following sections, agriculture can still feed the world without further harming the atmosphere.

LOCATION, LOCATION, LOCATION

As we discuss in Chapter 5, farmers are clearing large portions of the Amazon rainforest to make way for more soybeans and cattle. This deforestation is disastrous for the climate — by itself, it's about 2.2 percent of global GHG emissions. And the Amazon, known for years as the world's largest carbon sink, has become a carbon source. The most climate-friendly farms are situated on land that doesn't require cutting down rainforest. Meanwhile, both industrialized and developing countries are displacing land for agriculture in favor of development. Paving over land already in use for local agriculture often forces people to clear forests to plant their crops.

The world needs large-scale solutions to address deforestation for farmland. Governments in countries where deforestation for farmland is a major issue, such as Brazil, need to firmly regulate land use so that they can begin to deal with this major climate issue. We discuss how Brazil's dealing with deforestation in Chapter 12.

DEALING IN DIRTY SOLUTIONS

People often confuse soil with dirt. Not farmers. Still, sometimes, fighting climate change can be a dirty job. Healthy soil is a critical partner in absorbing carbon, but it can also be a major source of carbon dioxide emissions. If farmers modify farming practices to protect the health of the soil, they could cut back substantially agriculture's carbon dioxide emissions.

Here are some ways farmers can reduce GHG emissions:

>> **Change how they manage weeds so that they don't have to till the soil as often — if at all.** In short, the less farmers disturb the soil, the better. No-till agriculture increases carbon sequestration (refer to Chapter 2) in the soil.

>> **Use nitrogen fertilizer more carefully, which adds nitrogen dioxide into the atmosphere.** It also in the major contributor to ugly blooms of algae in freshwater (you've probably seen pictures of the green scum) that leads to death of lakes and streams. Careful use of this fertilizer means simply figuring out how much nitrogen you need to add so that you don't use more than necessary, which can help ensure that farming gives off fewer emissions.

>> **Move away from chemical fertilizers and explore greener alternatives, such as organic farming methods.** Producing and transporting chemical fertilizers, which are generally made from fossil fuels, is extremely energy-intensive. Studies have shown that organic farming methods use about half of the energy of conventional farming, and also sequester more carbon dioxide in the soil.

Reducing rice farms' emissions

Currently, land management among rice farmers is particularly poor and results in the emission of methane, a particularly potent GHG. In flooded rice fields, organic material breaks down, and because the still waters form an *anaerobic* (oxygen-free) environment, that breakdown results in the release of methane. Rice farmers can reduce these emissions by keeping the soil dry in the nongrowing season, adding any organic materials the soil needs during that dry period, rather than leaving the fields flooded to supply nutrients. In a dry field, *aerobic decomposition* (decomposition that occurs in an environment with oxygen) can happen, which doesn't produce methane.

GET A GREEN BELT BY PLANTING A TREE

Nobel Peace Prize–winner, the late Wangari Maathai created the Green Belt Movement, which is actively planting trees across mid-Africa. The problem in Wangari's eyes was simple: Climate change and related environmental problems could be traced to deforestation. The solution was just as clear: Plant trees — a lot of them.

Starting in the 1970s, she engaged women in a grassroots effort that led to planting more than 20 million trees. Her Kenya Green Belt Movement has since gone international, drawing in other African countries, including Uganda and Zimbabwe.

In 2007, at the UN Climate Change Conference in 2007, Maathai challenged the whole of the African continent to plant 1 billion trees. A true global hero and leader of the Green Party of Kenya, she died of cancer in 2011. Her work continues today. You can find out more about the initiative at http://greenbeltmovement.org.

THE GLOBAL-WARMING BEEF WITH COWS

One of agriculture's major sources of GHG emissions is livestock. With the global diet eating more meat, the world has an ever-increasing number of cows, which means an increase in methane. Cows are really gassy creatures — their digestive process (which makes it possible for them to break down the cellulose in grass so they can digest it) generates a lot of the potent GHG methane (sometimes called *natural gas*). When the cows belch, as they do a lot, this methane goes off into the atmosphere, making global warming worse.

Scientists around the world are currently exploring different ways to help cows cut this socially impolite and environmentally unfriendly habit. In the United Kingdom, research is underway to see whether adding garlic or seaweed to a cow's diet can help cut back on the gas. German scientists are testing a pill that traps and eliminates the gas in a cow's first stomach (they have four). In Australia, scientists are experimenting with transferring bacteria from kangaroos' stomachs to cows'. (Kangaroos have a similar digestive system and diet to cows, but they don't suffer from the same gastric unpleasantness.)

Chapter **15**

Activists without Borders: Nongovernmental Organizations

When you hear the word *activist*, you might think of angry marchers shouting slogans and hoisting protest signs — and you wouldn't be wrong. But you can be an activist in more than one way. Around the world, people have banded together into groups, determined to prevent a global warming disaster. Some of these groups do indeed hold rallies and stage sit-ins, but others are far more comfortable in boardrooms.

One of the most powerful forces in the world arises from people working together to improve society. Organizations in pursuit of social goals have been around for nearly 200 years, fighting against slavery, campaigning for the right of women to vote, and protecting the natural environment. Today, groups of this kind are called nongovernmental organizations, or simply NGOs.

The nongovernmental part of NGO is critical. Because these groups aren't tied to any government, they can be single-minded, focusing on one thing: for example, fighting global warming. In this chapter, we take a look at these organizations, how they're working hard to realize their goals, and how you can get involved.

Understanding What NGOs Do

If you enter either global warming or climate change, and then nongovernmental organization, into a search engine, you get thousands of hits, listing countless groups. The sheer quantity of NGOs might seem excessive; after all, aren't they all working for the same goal?

The number of NGOs indicates the enormity of the problem of global warming and just how much work needs to be done. Every NGO represents a different segment of the world's population affected by climate change, and each group tackles the problem differently. Some groups strive to increase awareness about the issues. Some stay on the periphery of society, aggressively trying to provoke people into action, and others work from within both government and industry, attempting to prompt progress.

The next sections describe the varied and global nature of citizen engagement, which is everything from small, volunteer, and grassroots to large, professional-ized, and well-funded. And citizen engagement is everything in between. You can find the right place for you.

Educating people

The scientists doing research on global warming aren't writing for a broad audi-ence; they're writing for their peers, to further scientific knowledge. This com-munication is extremely important because it ensures that science is constantly moving forward, building on new discoveries. Unfortunately, laypeople — the public at large — don't always hear about important research. That's where NGOs come in.

Making scientific information approachable

NGOs take complicated technical scientific reports, which are available to the gen-eral public, and translate them into language that the average person can under-stand. These people working as translators between science and the public are very familiar with the science and also understand how to communicate it in non-science terms. Many NGOs, such as the World Wildlife Fund (WWF) and

Greenpeace, write climate change reports that take key pieces from scientific reports and make them digestible to the reader. NGOs communicate via fact sheets, brochures, and websites, as well as at conferences and workshops, and even in material prepared for school curricula. These organizations hope that they can present the information in a way that motivates people to act.

As a result, NGOs fret over even the most essential terms to ensure that the words have the biggest possible impact. For instance, NGOs around the world struggled over what to call this human-made climate crisis. Global warming? Climate change? Most groups opted to call the threat "climate change" because some places will actually get colder when the global average temperature increases. U.S. groups went with the more popularized "global warming," a term with more immediate impact.

When translating science to lay language, NGOs need to be careful not to distort the facts. Many of the larger international environmental groups have scientists on staff. And, quite often, NGOs ask scientists to review their work to make sure it remains accurate. (Similarly, some of the leading scientists in the world reviewed this book to make sure we properly presented the science!)

Organizing and taking action

Groups like WWF, Greenpeace, and Sierra Club tackle many issues in conservation and protection of the environment and human health. As governments failed to act to meet the goals of climate agreements and to follow the warnings of scientists, new NGOs have emerged with a sole focus on climate action. Leading science and nature writer Bill McKibben founded 350.org, which refers to the atmospheric concentration of carbon dioxide (described in Chapter 2) — the 350 stands for 350 parts per million (ppm) of carbon dioxide in the atmosphere, the last really safe zone for carbon. At the time of this writing the Earth has hit 412 ppm, so 350.org holds a long-term vision of getting the concentration of carbon back down to 350 ppm (long term because global atmospheric concentrations won't go back down to 350 ppm — even with no burning of fossil fuels and even with massive efforts to pull existing carbon out of the atmosphere — for at least 100 years. (Refer to the sidebar, "Joining the climate change team(s)" later in this chapter.)

Keeping watch

NGOs play a key role as watchdogs over government and industry. They keep a close eye on what impact government and industry are having on climate change, and they're quick to point out when the powers that be don't live up to their green commitments.

These organizations are a big part of the reason that climate change issues make so many stories in the media — journalists often depend on organizations to share the top-hitting climate news story and connect them with the right people to talk to. (See Chapter 16 for more about the media.) Watchdog groups strive to ensure that the general public is as well informed as possible on climate change issues and what industry, government, and businesses are — or aren't — doing.

Organizations often publish report cards that score businesses or governments on their action on climate change — a clear way to communicate the state of the success of these actions, which the public might not otherwise know about.

WWF, for example, exposed Nike's high-emission running shoes and raised so much awareness about it that Nike agreed to change the gas used in the shoe air pockets, greatly reducing its greenhouse gas (GHG) emissions. Sinks Watch, as another example, is a project of the World Rainforest Movement, whose goal is to track and critically assess credits for offsetting through carbon sink projects. (Check out Chapter 11 for more about the carbon markets and offsetting.)

Getting the word out

Some NGOs have a reputation as rabble-rousers, constantly raising a ruckus. Greenpeace, for example, is famous (or is that infamous?) for unfurling enormous banners from high-rise buildings, off suspension bridges, or anywhere else their efforts can draw attention — even if it's against the law.

As the climate crisis worsens, new and more radical protests have emerged. In the 1990s and early 2000s, most people weren't seriously concerned that global warming could lead to mass extinctions including even us. It's not such a remote possibility now, as scientists agree that it's a tiny risk. So new groups have sprung up, among them Extinction Rebellion. Extinction Rebellion's tactics, blocking traffic and bridges, don't make them many friends beyond those already committed.

These groups know that their banners and blockades won't make an immediate difference. They're doing it for the guaranteed media coverage of their dramatic actions, which ensures that people will address the issue in a public forum.

REMEMBER

Whether hanging banners or organizing mass protests, these acts are most certainly attention grabbers. But not everyone loves these actions — not even everyone in the climate change awareness community is a fan. Some argue that the time these NGOs spend in these attention-drawing endeavors would be better spent working with industry or government to find solutions. Advocates for the stunts point out that direct actions bring a visual element to global warming, a threat that looms like a slow motion tsunami, which is hard for media to cover.

People blocking traffic ask drivers to think about whether it's better having delays now from protesters or having delays later from washed-out roads and mudslides if societies and governments fail to act to effectively protect a stable planet.

Working with industry and government

From much of the news coverage you see, you may think that all environmental groups spend their time yelling at corporations and governments across barricades — and that industry and government spend their time trying to build bigger barricades. That's not actually the case (although that tends to be the kind of conflict that makes news). A very large number of NGOs make progress by working with industry and government as we discuss here.

Corporate cooperation

Environmental group representatives are often asked to sit on advisory panels for industry and business. These panels consist of people who offer companies outside perspectives on their plans and actions. For environmental activists, being on advisory panels presents opportunities to help companies develop a greater understanding of the ecological impact of their actions.

The benefits of NGO–corporate partnerships go beyond advisory panels, however. Businesses are a link for NGOs to create policy changes, and the support of an NGO can help endorse the climate-friendly practices of a business. With their comprehensive understanding of the causes of climate change, NGOs can offer industries real insight into problematic practices that have been causing major emissions. And businesses can often find practical (and economical) ways to implement GHG-reducing technologies and strategies. By bringing these two specializations (so to speak) together, industries can get a more holistic perspective on best practices for business and industry operations.

GOOD NEWS

Some partnerships between NGOs and businesses, such as the following, have really helped reduce GHGs:

>> The Climate Group is an international organization made up of companies and government representatives to advance leadership on climate change, work with partners such as HSBC Holdings and Intel to improve management systems, save energy, set targets for reducing emissions, and make a profit while they're at it. The Climate Group has played a catalytic role with governments in launching the Below 2 Degrees Coalition. Below 2 Degrees Coalition has members entirely from governments — at the state and municipal level. They work together to move other governments to bring in tougher climate action.

» The Pembina Institute in Alberta, Canada, has worked with companies such as Suncor in the oil industry to reduce their emissions. Producing usable energy from the oil sands is energy-intensive in its own right — generating about 8 percent of Canada's total emissions.

» WWF leads a project entitled Climate Savers, engaging companies such as IBM and Lafarge to reduce their carbon dioxide emissions. WWF has engaged with dozens of major corporate partners to press for 1.5 degrees C. The Climate Savers have developed much more critical tools to differentiate between nice words form business and real measurable progress.

» The Business Environmental Leadership Council (BELC) is a U.S.-based corporate network. Founded in 1998, it has grown to now include 38 mostly Fortune 500 companies working to reduce GHGs. The BELC was very active at COP26 in Glasgow pressing for conclusion on the carbon market (article 6 of Paris Agreement) to avoid scams like *double counting* (including the same emissions reductions in more than one measurement). Refer to Chapter 10 for more information about the carbon market and problems like this. With combined revenues of nearly $3 trillion, the organization has clout.

Government action

On national and international levels, NGOs are working with governments to fight climate change. In individual countries, many governments recognize and often draw on the expertise of NGOs — despite the fact that such groups often disagree with government policy and are working to change it.

REMEMBER

Governments and NGOs are naturally complementary to one another. NGOs are generally created to fill a gap that people have identified in government. Both the government and NGOs have a mandate to act in the best interest of the public, but NGOs focus on particular issues. NGOs aim to bring the voice of the public to the government.

Internationally, large NGOs are very active, participating in every climate negotiation since the first meetings toward a global treaty began in 1990. The United Nations (UN) recognizes credible and well-organized NGOs as observers to global negotiations. The role of *observer* isn't nearly as passive as the name may suggest; observers can speak at meetings, but they can't vote on negotiated text for treaties.

In their speeches at these conferences, NGOs can ensure that the grassroots are represented and that the issues aren't lost in any political maneuvering. They also meet regularly with the delegations of governments to discuss policy and political stances. To bring awareness to the public, NGOs communicate with reporters and greatly help focus the attention of the media on key issues. The collective work of these NGOs helps move the negotiations forward and puts pressure on the politicians at the table.

BANDING TOGETHER

From the earliest stages, the nongovernmental organization (NGO) community realized the climate crisis would require a global solution. So, the NGOs formed their own global network, the Climate Action Network (CAN), which comprises more than 1800 NGOs from more than 130 nations. Through CAN, NGOs share information and strategies to push for progress internationally, nationally, and regionally.

You can check out CAN's website at www.climatenetwork.org. Individual member countries have Web sites, too, with useful links and resources:

- **Australia:** www.cana.net.au
- **Canada:** https://climatereactionnetwork.ca
- **United States:** www.usclimatenetwork.org
- **Western Europe:** https://caneurope.org

Indigenous leadership

Increasingly at local, national, and international levels, the voices of Indigenous peoples and Indigenous climate organizations are playing a key role in climate action. In the Paris Agreement, the critical role of Indigenous peoples is referenced — as is the shift in world view to recognize "Mother Earth."

Unlike NGOs, Indigenous peoples are often recognized as having sovereignty — as nations within other nations. The negotiation of the United Nations Declaration on the Rights of Indigenous People (UNDRIP) has fortified the global recognition of the unique place of Indigenous peoples in relationship to the governments of the world. Their voices can't be lumped in with NGOs.

At international COPs, Indigenous voices are heard in the plenary sessions — if only for a few minutes and rarely. Many governments, such as the United States and Canada, include Indigenous representatives on the government delegations.

Outside the halls of COP, Indigenous peoples are heard loudly. In Glasgow at COP26, Indigenous peoples from the Amazon, British Columbia, and the northlands of Scandinavia were heard in powerful speeches to 150,000 people.

Indigenous people has been effective in blocking new fossil fuel infrastructure at Standing Rock, North Dakota, on Wet'suwet'en Territory in interior British Columbia, and in the Achuar territory of Ecuador. One study found that Indigenous actions around the world had been more successful in keeping fossil fuels in the ground than governments. These actions are inherently political.

INDIGENOUS LAND DEFENDERS

Colonizers of hundreds of years ago took land from the original peoples of the Americas. The land was sometimes taken in bloody conquest, other times through negotiated treaties. However, treaties with the original peoples of North America have been violated more than they were respected. As a result, there have been modern conflicts over justice and rights.

Furthermore, there's a conflict of world views even though many Indigenous peoples have shifted to the mainstream world views and are comfortable in corporate boardrooms or in developing projects that exploit natural resources in modern ways. Still, traditional Indigenous world views have survived despite centuries of oppression. Oversimplifying, that traditional view sees the world in circles of relationships. Indigenous peoples in many parts of the world see the nonhuman life of this planet as relatives. Traditional values approach the taking of an animal for food, for example, as an act that must be imbued with respect, reverence, and gratitude. Nothing must be wasted. As a result, Indigenous world views are grounded in abundance.

In contrast, European world views are more linear than circular. The Judeo-Christian tradition stemming from Genesis tends to see nonhuman life as here for humanity's use. Although many religious scholars point out the relationship from Genesis was about stewardship and responsibility, the European or settler culture world view sees human beings as "top dog." That world view is also grounded in scarcity — humans are always worried they'll run out of something. So we keep more than we need, we waste a lot, and as a result, there is scarcity. York University professor Peter Timmerman pointed out the irony that those who believe in a world view of scarcity create the scarcity they most fear. The Indigenous world view places abundance at the center — and creates abundance.

Given these historical and philosophical conflicts, no wonder many Indigenous peoples around the world see the climate crisis as far more urgent. Humanity is literally ripping Mother Earth apart.

Meeting This Generation

People under 15 years of age make up one quarter of the world's population. That's a big share, and because many of the major climate changes are projected to hit 20 years from now, youth groups have a special role in the climate change NGO world. Youth organizations can make a difference by educating their parents and larger community, by greening university campuses, by pressing governments for change, and by promoting a low-carbon lifestyle.

The importance of youth

Youth play an important and unique role. They can add a real sense of urgency to climate talks by stressing that the future — which may seem abstract to policy-makers, industry, and the public — is very real for them. When scientists say, "This is what the world is going to be like in the year 2050," this young generation will live through those major changes. The threats are very real, and so is the need to immediately implement solutions.

The most useful element that youth bring to the table is optimism and high energy. They also invoke a sense of moral obligation in their elders. The involvement is genuine and fresh, and the lack of years of experience means that most youth see climate change with a sense of simplicity. When youth speak about global warming, they do so without employing numbing jargon or invoking complex political issues. What matters to them is their future, and they say so purely and simply.

Groups that speak up

In just the past handful of years, coalitions and networks of youth organizations working on climate change issues have formed. They're sharing resources, organizing networking meetings between youth climate groups, and bringing youth representatives to UN climate change conferences. Here are just a few of those groups:

» **Fridays for the Future** (https://fridaysforfuture.org/): Swedish school girl Greta Thunberg founded Fridays for the Future. Her story is amazing: She was just 15 years old, one student, taking one small action — just sitting in front of her school in Stockholm on Fridays with a handmade sign. Presidents and world leaders of all kinds now meet with Greta and try to convince her that they will take the necessary action. She isn't easily fooled and has described those politicians who claim to be acting of just saying "blah, blah, blah." You can find Fridays for the Future groups in more than 200 countries around the world.

» **The Energy Action Coalition** (www.energyactioncoalition.org): This group brings together almost 50 youth-run organizations across the U.S. and Canada. Not restricted to climate issues, the coalition has an array of foci, ranging from environmental justice and politics to community and education. It runs the Campus Climate Challenge program, which implements clean energy policies and emission reduction strategies at colleges and universities while raising awareness among students. This program already has more than 570 campus groups involved. The Coalition was the brainchild of Billy Parish, who dropped out of Yale to do more important things, like save the planet.

CLIMATE JUSTICE – NGOs THAT HEAR THE VOICES OF THE POOR AND GLOBAL SOUTH

As the climate crisis has worsened, there is a growing awareness of the unfairness of the impacts. As we discuss in Chapter 12 the impacts of the crisis are far worse for the people who did the least to create the problem — the poor everywhere and the peoples of the developing world — often referred to as the "Global South."

These issues are now part of the main talks at COPs. The term "climate justice" is included in the preamble of the Paris Agreement. It was never part of earlier agreements.

More and more NGOs focus on *climate justice,* fighting against the inequity of the problem and not just the problem itself. Existing organizations that worked against poverty around the world — like Oxfam, many church relief organizations, and CARE — are also part of the climate movement. They work at COPs and in the time in between to rally for climate financing.

Getting Involved

Because many NGOs rely on volunteers, they're always looking for people to join the team. If you're looking to get involved in climate change issues, participating in an NGO is one of the best things you can do. You can make a difference, especially as part of a larger group. As a member of an organization, you can rely on a support network of people who have experience addressing climate change issues; they can help you find where you can best contribute.

Remember the classic American poster for the war effort in the Second World War: "Uncle Sam wants you!" You might now imagine a new one: "Mother Earth wants you!" The next sections tell you where you can sign up.

Seeking out groups

As we show in the section "Understanding What NGOs Do," earlier in this chapter, a wide array of NGOs exist. One's bound to match your interests. Here are some suggestions about how to find a group that's right for you:

>> **Attend a conference or event.** These hotspots for meeting representatives from NGOs can help you find out about each group's projects and how you can get involved.

>> **Contact your area's environment department.** Often, your provincial, state, or federal government's department of the environment has lists of organizations working on climate change. (We share some government websites that can point you in the right direction in Chapter 21.)

>> **Show up at Green Drinks.** Starting off as a good idea by a small group of friends in 1989 in the U.K., the event Green Drinks now occurs monthly in more than 30 countries and in more than 350 towns and cities around the world. People from academia, business, government, and NGOs who are working or interested in the environmental field meet up. It's a great place to meet the inner circle of climate change–savvy people in your area and get connected with groups acting on the problem. Green Drinks is widespread across the United States, the United Kingdom, Australia, and Canada. Check out www.greendrinks.org for more.

>> **Go online.** The sidebar "Joining the climate change team(s)," in this chapter, lists the websites of some of the major NGOs. Many of these larger organizations have local chapters or can at least direct you to groups in your area.

Helping out

Organizations often have multiple campaigns going on at the same time and always appreciate help at any level. You can get involved in many ways, depending on your skills and interest.

Most organizations have Get Involved or What You Can Do sections of their websites that link directly to current projects for which they need help. Here are some examples of volunteer positions:

>> **Fundraisers:** Raising money for organizations working on climate change ensures that they have funding to keep fighting climate change.

>> **Organizers:** If you have strong organizational skills, many groups will be more than happy to have you help organize (or even take on organizing) public awareness events or conferences on climate change in your community.

>> **Influencers:** Use Facebook, Twitter, Instagram, and other social media platforms.

>> **Public speakers:** Speaking at events, big or small, is a help to organizations. Often, organizations offer training, so that you know just what to say.

>> **Writers:** Organizations are always on the lookout for people to write letters to government, business, and industry representatives. Writers are also a big help in creating press releases and articles to get issues into the public eye through the media.

JOINING THE CLIMATE CHANGE TEAM(S)

Many great groups are working on climate change issues. Some focus only on how to make religious buildings more energy efficient, some concentrate on greening schools. Some are active on a local scale in their own community or region, and others are taking it global. Here's a tiny snapshot of major NGOs in Australia, Canada, the United Kingdom, and the United States that focus on climate change:

- **350.org (https://350.org):** Active since 2007, 350.org has taken up successful divestment campaigns to pressure large institutions, particularly universities to remove fossil fuel investments from their portfolios. Divestment was a significant part of the campaign to end apartheid. Billions of dollars in fossil fuel investment have been dumped thanks to 350. 350.org is now active all around the world.

- **American Solar Energy Society (www.ases.org):** Active since 1954, this organization promotes and implements the use of solar energy, energy efficiency, and other renewable energy technologies. You can join a local chapter and receive guidance on installing solar energy in your home.

- **Rotary International (www.rotary.org):** Are you a Rotarian? This effective globally active service organization has now taken on the environment and climate as a priority area of work. The Environmental Sustainability Rotary Action Group (ESRAG) is an exciting new development in the global climate movement. Check and see if there is already a Rotary Club in your community and tell them you want to join to work with ESRAG.

- **David Suzuki Foundation (www.davidsuzuki.org):** A large Canadian organization founded by icon environmentalist David Suzuki, this group offers a Nature Challenge to members to reduce their carbon footprint and work together to reduce our emissions.

- **Friends of the Earth (www.foei.org/):** This organization has more than 2 million members and 5,000 local chapters in 69 countries. Members volunteer locally with chapters to reduce the GHGGHG emissions of their communities, and many members also get involved by responding to requests from the Friends of the Earth headquarters — such as writing to their government representatives on an urgent climate-related issue.

- **Greenpeace (`www.greenpeace.org/~climate`):** Greenpeace is a major international organization with many climate change awareness programs: You can sign up for a seven-step climate change challenge, or you can become a Cyber Activist — receiving online alerts of when and how to take action. Greenpeace accepts all kinds of volunteers; they work with the skills you have and fit you where you can be most effective.

- **Sierra Club (`www.sierraclub.org`):** What began in 1892 as a club dedicated to bringing people into the great outdoors to experience the wonder of nature has grown into one of the strongest environmental groups and advocates for battling climate change. You can join in Canada and the United States and belong to a local chapter or youth branch, go on outdoor outings, and volunteer for campaigns. You can find Sierra Club Canada at `www.sierraclub.ca`.

- **Union of Concerned Scientists (`www.ucsusa.org`):** This organization, founded and staffed by scientists, promotes a high level of public understanding and practical science-based solutions towards a better world. Anyone can join and contribute to the discussions and research that shape the group's reports and policy recommendations. The UCS has been active for more than 50 years throughout the United States, addressing critical issues, such as climate change.

- **World Wildlife Fund (`www.worldwildlife.org/climate`):** WWF is active all over the world. When you become a member, WWF continually updates you on WWF climate campaigns and how you can get involved in international climate conferences, gives tips for making green changes at home, and offers partnerships to green your business. If you live near a WWF office, you can volunteer on-site, too.

» Looking at how Hollywood and its stars are responding to climate change

» Checking out climate change blogs

» Curling up with a few climate change books

Chapter **16**

Lights, Camera, Action: The Media and Climate Change

Not too long ago, climate change was the exclusive domain of climatologists and environmentalists. Now that its effects are undeniable, it's on everyone's lips and getting serious exposure in popular culture. This is good news, as far as we're concerned — the more people know about climate change, the greater the momentum for change.

But is any publicity good publicity? In this chapter, we take a close look at how the news and entertainment industries cover global warming and show how the science that we cover in Part 1 sometimes gets lost in the push for a good story.

Growing News Coverage

Once relegated to articles in science journals and the back pages of the newspaper, climate change now merits cover stories in mainstream magazines, front-page stories in the papers, and feature reports on news broadcasts.

Although the number of climate change stories is increasing worldwide, the rate of growth is different from nation to nation. The question of quality and quantity of media coverage has been an area of academic interest. In response, a center, the Media and Climate Change Observatory (MeCCO) based at the University of Colorado, monitors 126 news sources in 58 countries, studying the level and accuracy of media coverage of climate.

Without this coverage, public awareness of climate change wouldn't be where it is today. But, as we cover in the following sections, you need to read those feature stories and watch those reports carefully, no matter where they're published or broadcast, to ensure you don't get misinformed.

Bias and balance: Distorting the story

In 2004, academics first noted that media tend to give too much attention to the few scientists who don't agree that climate change is serious and caused by humans. Because journalists are supposed to give both sides of a story, sometimes they actually create a bias in their reporting. One of the first to study this were academics Jules and Maxwell Boykoff. Their study of more than 600 randomly selected articles on climate change that appeared in *The Wall Street Journal*, *The New York Times*, *Los Angeles Times*, and *The Washington Post* found that coverage could be described as balanced.

Although this "balanced" reporting might seem fair, the likelihood that humans are contributing to climate change was then 95 percent certain, according to the Intergovernmental Panel on Climate Change (IPCC). Since then, climate change has become 99 percent certain. Yet many more studies have been conducted on media "balance." They all find that trying to hear both sides — even when one is a dwindling minority — creates the inaccurate perception that it's an equally weighted debate. One of the most recent studies by Michael Bruggemann and Sven Engesser found that even as the scientific consensus grows, more individuals in the United States and China refuse to accept the science. Researchers tie what they call "entrenched denialism" to the way journalists cover the climate crisis.

Consider the source: Being an informed media consumer

You sometimes have to really work to figure out when something you read or see in the media is accurate. But by being a careful media consumer, you can tell the difference between credible information and inaccurate reporting. Watch out for these key issues when reading articles or watching features about global warming:

>> **Questionable experts:** Television news reports need talking heads, and newspaper articles require people who can act as sources — and plenty of people are willing to offer themselves as experts, with or without the knowledge required. A handful of "experts" always support the climate denier side of the argument. These pundits like media coverage and appear to have good credentials, but they may not actually do research on climate, they may receive funding from fossil fuel companies, or they may distort the work of others. Media-friendly experts who stress immediate action on global warming can be problematic, too. Although they feel passionately about the issue, these individuals may communicate the wrong information by oversimplifying complex science or exaggerating global warming's effects.

TIP

You can tell whether or not these people are experts by checking to see whether the news story references any scientific research that the expert has actually had published in a peer-reviewed journal.

>> **Old research or evidence:** If the report doesn't mention the date of the evidence it's citing, that's a red flag — the evidence could be an old theory that has since been refuted. For example, the documentary *Planet of the Humans* focused on the effectiveness of renewable energy using statistics from the 1980s. Since then, wind and solar have made enormous strides, but filmmakers trying to demonize renewable energy conveniently ignored that data. (See the following section for more about this documentary.)

>> **Source of information:** If a report references the IPCC, you can trust that information — as long as the writer or producer hasn't misunderstood the words of the IPCC. A scientist who has published and peer-reviewed articles offers another good source of reliable information.

Blogs and Wikipedia, on the other hand, are generally the least credible sources of information about climate change. Anyone can create or contribute to these websites. (Nevertheless, we look at a few of the most popular blogs in the section "Worldwide Warming: Climate Change Blogs," later in this chapter.)

TIP

For reliable online information, try government, nongovernmental organization (NGO), and university websites. (We recommend some of our favorite resources in Chapter 21.)

>> **Oversimplification:** News reporters often simplify information to a point where it's actually incorrect or overexaggerated. For example, if a scientist says sea levels could rise by 13 to 16 feet (4 to 5 meters) if the Western Antarctic Ice Sheet collapses, a reporter might say the scientist has *predicted* this will happen. Predicting and projecting are two different things. A *prediction* is an assertion that something *will* happen, but a *projection* is an extrapolation of future events if current trends hold true. Change the trend lines and the projection will change.

Complex scientific issues don't fare well in the land of the soundbite. If the article you're reading or the report you're watching offers extreme outcomes, check for further background information to ensure you get the whole story.

>> **Surveys:** Surveys are a great way to gauge public opinion — if they're large enough, if the questions are well framed and don't suggest the answers, and if the methodology is scientific. Sometimes those conditions are met, but often news media will report on parts of polls as if they were as accurate as the bigger surveys. When reading about a poll, check especially how many people were surveyed. A small number of people (say, 50) means that the margin of error can be very large and the poll or survey results probably don't reflect the greater population.

Focusing on Science on the Red Carpet

Climate change is so hot that Hollywood studios haven taken notice for quite some time. And their stars are acting, as well — not just on-screen, but in their personal lives. But what's the real story? How much truth is there in *The Day After Tomorrow*? How green — really — are those jet-setting stars? And can George Clooney make a difference by tooling around Tinseltown in an electric car? These sections examine Hollywood's depiction of climate change — both the good and bad.

Movies: Facts and (science) fiction

Although climate change is constantly gaining public attention, it hasn't served as the subject (or even the backdrop) for many films. Because climate change poses a complex problem that requires complex solutions, it doesn't make for the stuff of Hollywood fodder. For the most part, when climate change has turned up on the silver screen, its portrayal has been sensationalistic, setting the scene for science-fiction dystopias. Thankfully, some documentaries set the story straight, and some of the best work that's being done on the screen is explaining the problem to kids, who will be on the front lines of climate change impact while they grow up. We examine both here.

Silver screen stories

Perhaps the classic fictional film that addresses global warming, *The Day After Tomorrow* (2004), is a well-intentioned movie that's scientifically credible . . . for the first 15 minutes. The movie opens promisingly — the film's hero, a climate scientist played by Dennis Quaid, briefs a world conference on his theory: When

Arctic ice melts, the freshwater released will interfere with the oceans' thermohaline currents, slowing them down and leading to climatic disasters. (Chapter 7 discusses the actual computer modeling of this possibility.) The fictional U.S. vice president reacts realistically, demanding to know who would pay the costs of Kyoto (see Chapter 11).

But after this believable beginning, the plot becomes completely driven by special effects, and the real science is lost in the (literal) deluge. We can't deny the coolness of watching freak tornadoes obliterate Los Angeles, snow pummel New Delhi, and killer hail devastate Tokyo — but by the time a deep freeze destroys the entire northern half of the planet, we're no longer noting scientific accuracy. By then Dennis Quaid's character has set out on an epic trek to save his son, played by Jake Gyllenhaal. Although special effects dominate the movie, the producers acknowledge the liberties taken and include useful interviews with scientists in the DVD's special features.

Science fiction films have dominated the list of fictional movies that deal with global warming. In terms of quality, they're a mixed bag:

- » **Waterworld (1995):** Kevin Costner starred in the ill-fated *Waterworld* as a mariner sailing the Earth after global warming triggered flooding, which has covered almost all land. The only ground still above water, called Dryland, turns out to be the summit of Mount Everest. Sadly for the movie's credibility, the world doesn't have enough water to flood the whole thing. The highest estimates of how far sea levels could rise from global warming max out at about 650 feet (200 meters). That's a lot, but the Waterworld scenario would need something like 30 times that depth.

- » **A.I.: Artificial Intelligence (2001):** This post-disaster film takes place after all the polar ice caps have melted and most coastal cities are entirely flooded, pushing civilization inland. Although humanity might never see a robot with feelings, which is the subject of Steven Spielberg's film, flooding of low-lying cities is a realistic possibility. If people keep producing emissions at the current rate, coastal waters could rise 8 to 24 inches (20 to 60 centimeters) by the year 2100.

- » **Age of Stupid (2009):** Although fictional, this powerful film set in 2055 has elements make it feel like a documentary. The lead, Pete Postlethwaite, plays a lone survivor of climate disaster. He inhabits a futuristic archive with access to vast footage of what occurred before the collapse of human civilization. He pursues the question "Why didn't we save ourselves when we still had the chance?" It's highly accurate and scientific, although if humanity acts on climate change now, the year 2055 won't be a time of post-apocalyptic grief.

- » **Don't Look Up! (2021):** After a long drought of Hollywood flicks about climate, Netflix launched a blockbuster in late 2021. Director Adam MacKay also wrote the screenplay based on a story by David Sirota. The film features an A-list of stars: Meryl Streep, Leonardo DiCaprio, Jennifer Lawrence, Cate Blanchett, Jonah Hill, Mark Rylance, Rob Morgan, and more. The plot line doesn't directly involve climate change. Two astrophysicists discover a giant meteor hurtling toward Earth called a climate-killing comet. The movie is about what happens as they try to convince everyone of the severity of what's coming.

Everyone sees it as satire about how the worlds of politics and media are able to ignore a growing threat to survival. Climate scientists from around the world responded positively, seeing their own real-life experience in the fictitious scientists trying to persuade politicians and reporters of the growing threat and to take action to save Earth. Film reviewers, who naturally focus on its value as entertainment, were more divided. More to the point, it has, as intended, provoked a lot of thought and started a lot of conversations.

Documentaries

The first influential full-length movie on climate change was the documentary *An Inconvenient Truth* (2006), featuring former U.S. Vice President Al Gore. The film walked away with the Academy Award for best feature-length documentary.

An Inconvenient Truth is based on Gore's lecture presentations that highlight big, catastrophic events that climate change may cause. Called a true horror film by some for the impact it has on viewers, it makes the scale of the problem clear to audiences — that the situation is way beyond such easy fixes as switching light bulbs. However, it does outline what individuals can do, and it definitely motivated many viewers. The movie essentially changed global thinking, moving climate change to the forefront of many people's minds. Gore himself, along with the Intergovernmental Panel on Climate Change (IPCC), was awarded the Nobel Peace Prize in 2007 for his climate change work.

Since the success of *An Inconvenient Truth*, a large number of accurate and riveting climate documentaries have inspected the climate crisis. Here are some documentaries that tackle the subject well:

>> *Chasing Ice* (2012): Filmmaker Jeff Orlowski traces the real-life work of scientist James Balog and his team of researchers conduct the Extreme Ice Survey. Dramatic footage brings views up close and personal to massive ice collapse events.

>> *This Changes Everything* (2015): Based on Naomi Klein's book of the same name, this film takes a broad sweep at global activism in response to the climate emergency. Directed by Avi Lewis, Klein's partner, the human element in this film is how the work to protect climate has impacted people from around the world.

>> **The Anthropocene: The Human Epoch (2018):** If people are the focus of the other documentaries, the images and photographs of award-winning artist Edward Burtynsky are the stars of this film. Burtynsky's collaborators in making this film, Jennifer Baichwal and Nicholas de Pencier, have created a powerful work of art that speaks for itself.

>> **Kiss the Ground (2020):** Highly recommended is this film featuring Woody Harrelson. The message of the film, ground in solid science, is that by doing farming differently a lot of carbon can be pulled from the atmosphere.

>> **I Am Greta (2020):** This film is the story of the amazing Swedish school girl, Greta Thunberg, who has become one of the most influential climate voices in the world. This film has appearances from an assortment of people, including Catholic Pope Francis and U.N. Secretary General Antonio Guterres and is an engaging look at an extraordinary — and unlikely — heroine. Thunberg and her organization Fridays for the Future have managed to mobilize millions to climate strikes- national and global.

WARNING

Filmmaker Michael Moore lent his name to a poorly researched documentary called *Planet of the Humans* in 2020. It tries to make a case that environmentalists are proposing climate solutions that make climate change worse. The argument is that cutting down forests to make biofuel is supported by climate activists. It isn't. And it uses out-of-date statistics to suggest that solar energy and wind power are wasteful and ineffective. Fortunately, experts provided critiques to set the record straight. Unfortunately, millions watched it online and it is hard to know how many accessed the corrections.

For the kids (and adults, too)

You may think that climate change is unlikely subject matter for children's entertainment, but a couple of recent kids' movies have tackled climate change:

>> **Ice Age 2: The Meltdown (2006):** This movie makes good-humored references to the natural Ice Age and warming cycles while touching on such current threats as flooding. Roger Ebert noted, "If kids have been indifferent to global warming up until now, this *Ice Age* sequel will change that forever."

>> **Happy Feet (2006):** This environmentally minded movie features penguins who are dealing with the consequences of civilization's encroachment on the natural world. Although the movie doesn't expressly address global warming, it does get kids thinking about the impact of society's actions on the greater environment.

Following the stars

"Well, if Leonardo DiCaprio is doing it . . ." Okay, maybe you need more motivation than that to get involved, but the power of the stars is undeniable. Many celebrities are using their high profiles to advance green living and advocate action on climate change. DiCaprio, Arnold Schwarzenegger, Jane Fonda, Mark Ruffalo, and others are speaking out, getting involved with climate change organizations, and practicing what they preach by making smart choices about cars, travel, and home retrofits. Not everyone can spend hundreds of thousands of dollars on high-tech goodies, but everyone can do their part, and these stars deserve credit for making these choices and encouraging others to do the same.

GOOD
NEWS

The organizations that stars work with get a lot of press coverage and free publicity. Concerns exist about turning an urgent issue such as climate change into a pet cause — after all, a celebrity could be ill-informed, and the media would print their words anyway — but the benefits far outweigh any concerns. Most of the time, the stars know what they're talking about, and they're genuinely concerned about the issue. In this kind of situation, the ideals associated with celebrities (as they are so often in the media limelight) can influence people — and leaders — for the better.

Leading lights

Here are just a few of the stars putting their celebrity to work:

>> **George Clooney:** Twice named the Sexiest Man Alive by I magazine, he just got sexier. He drives the ultra-mini, one-seat-wide Commuter Car Tango and the Tesla Roadster, a top-of-the-line electric beauty.

>> **Leonardo DiCaprio:** Climate change is a passion for this actor, who co-wrote and produced an environmental documentary, *The 11th Hour,* and stars in *Don't Look Up.* He also created the Leonardo DiCaprio Foundation in 1998 to support a variety of environmental causes. DiCaprio sits on the boards of the National Resources Defense Council and Global Green USA. In his personal life, he has installed solar panels on the roof of his house.

>> **Willie Nelson:** Following up his foray into marketing "BioWillie," a blend of 20 percent biodiesel and 80 percent regular diesel, the veteran country singer has thrown his support behind the nonprofit Sustainable Biodiesel Alliance.

>> **Julia Roberts:** The *Pretty Woman* star uses environmentally friendly products (including her kids' diapers), has solar panels on the roof of her home, and uses biofuel in tractors and equipment at the family ranch. She's a spokesperson for Earth Biofuels, Inc., and chairs its Advisory Board. She's also involved in a community project to protect 100,000 acres of wildlife habitat in New Mexico's Valle Vidal.

>> **Arnold Schwarzenegger:** Famous for his role as *The Terminator,* Schwarzenegger quit his acting job to run successfully for governor of California. He's been a tireless advocate for action on climate change. (Check out Chapters 10 and 19 for some of his successes.)

>> **Robert Redford:** Redford created the annual Sundance Film Festival, where *An Inconvenient Truth* had its world premiere, and the Sundance Channel, which broadcasts a three-hour, prime-time programming block titled *The Green,* devoted entirely to the environment. Actively fighting for conservation for decades, Redford has been described as a steadfast, well-informed voice on environmental issues.

Carbon offsetting and green travel with the jet set

One area where celebrities don't often set a great example, however, is in their jet-setting ways. Some environmentalists give performers a hard time for the amount they travel by plane (often private), and for the extensive energy and fuel needed for major press junkets, music tours, and personal trips. Fortunately, many celebrities are beginning to carbon offset their air and ground travels — and the list is long, ranging from Tom Cruise to Nicole Kidman to Steven Spielberg. (Check out Chapter 6 for information about the real value of carbon offsetting.)

Although performing artists still tour, many of them try to make up for it along the way by fueling their buses with biodiesel blends, carbon offsetting, and providing information on environmental issues to their fans. The leaders in this movement include Bonnie Raitt, Willie Nelson, Jack Johnson, Alanis Morissette, the Barenaked Ladies, and the Dave Matthews Band. In fact, the Barenaked Ladies have their own carbon-offset organization, dubbed Barenaked Planet.

Worldwide Warming: Climate Change Blogs

Blogs continue to boom. A *blog* is an unofficial article or opinion piece posted by an individual online. Most blogs aim to spur back-and-forth discussion among readers and with the author.

Here are a few of the most-read, climate-centered blog sites:

- >> **Desmog Blog** (http://desmogblog.com): This blog hits on a wide range of climate change issues with the goal of removing the "smog" of media hype.

- >> **It's Getting Hot In Here — Dispatches from the Youth Climate Movement** (www.itsgettinghotinhere.org): The posts on this relatively new blog come from young people, from students to activists to professionals, posting on climate issues (mostly in the United States and Canada).

- >> **Real Climate** (www.realclimate.org): Although some of the language on this blog is science-based, the climate scientists who post here are writing for the public. They aim to stick to scientific discussion and stay out of politics and economic issues.

WARNING

Blogs can often be the source of the *least* credible information. You have to judge the legitimacy of the writing, which depends entirely on the author, their background, and their research. Remember, anyone with access to a computer can post a blog or comment on one.

TIP

When you're in doubt about a blog's credibility, click on the About Us link that most sites feature. The bio page that appears often reveals just how legitimate the blog's information is. If the individual or group has a clear bias, you can probably guess the influence that bias has on the information you're reading.

Bestselling Books: Reading between the Lines

The number of books on climate change has exploded in recent years. From science to fiction, fear to solutions, and children's books to adult titles, you have plenty to pick from. We discuss some of our favorites here.

True stories

A number of extremely well-written, nonfiction books on climate change have been published in the past decade or so. Some are long on the problem and short on offering solutions, which can be . . . well, a little depressing. Among the better-known titles are the following:

- >> *The Citizen's Guide to Climate Success* by Mark Jaccard (Cambridge University Press): Jaccard is an economist at Simon Fraser University in

British Columbia. He has been engaged for years with the economics, politics, and regulation of energy and climate change. He argues that people pressing governments for rapid action need always to take into account what is politically possible, and that leads him to argue for "flexible regulations" (like the successes shown in California) as the prime government tool, in addition to carbon pricing.

» *Electrify* **by Saul Griffith (MIT Press):** Griffith is an inventor, entrepreneur, and recipient of the MacArthur Fellowship Genius Award. He shows that rapid electrification of the entire U.S. energy system is possible in time to keep global temperature rise below within the IPCC's limit. His data shows that the "electrification of almost everything" is much less expensive for citizens and governments than carrying on with fossil fuels, and he calls for innovative approaches to financing the massive transformation.

» *Field Notes from a Catastrophe* **by Elizabeth Kolbert (Bloomsbury):** Based on Kolbert's articles for *The New Yorker,* this is a first-person journalistic look at the science behind and the impacts of climate change. Kolbert offers a well-researched and clearly explained account of the urgency of climate change while linking this to real places and stories from around the world.

» *Heat* **by George Monbiot (South End Press):** Monbiot, one of the U.K.'s most respected journalists, offers a truly radical approach to avoiding atmospheric tipping points. His writing reflects the immediacy of the climate crisis by demanding changes such as rationing energy use. Monbiot doesn't believe that the range of policy options we describe in Part 4 can get greenhouse gas (GHG) levels low enough fast enough. He could well be right. (See Chapter 19 for more about Monbiot.)

» *An Inconvenient Truth* **by Al Gore (Rodale Books):** Tied to the film of the same name (which we discuss in the section "Focusing on Science on the Red Carpet," earlier in this chapter), this current affairs book is the most accessible title on climate change currently available. Gore conveys his message with minimal text, using easy-to-read graphs to show the science behind climate change.

» *Values* **by Mark Carney (Signal, Penguin Random House):** Carney is a past Governor of the Bank of Canada and the Bank of England, and is now the UN Special Envoy on Climate Action and Finance. Carney says that the world is now facing the "tragedy of the horizon," by which he means that the value systems of governments and industries only look a short way out to the future, and aren't compatible with our deep human values because they don't incorporate into their decisions anything beyond their own short time horizons. He argues that the financial markets are already moving away from fossil fuels, that the call to "keep it in the ground" to meet the carbon budget can be realized, and that it's possible to keep to 1.5, only if the world gets back to working for the common good and for the values of the future.

- » ***The Weather Makers*** **by Tim Flannery (Atlantic Monthly Press):** Beautifully written, this book covers environmental science and issues in detail and depth. Though the content is a little overwhelming, Flannery offers solutions and a vivid writing style that draws you in. This book is a little frightening, but offers solid and entirely correct scientific information.

- » ***The Winds of Change*** **by Eugene Linden (Simon & Schuster):** Based in science and history, this book covers such overarching topics as the Gulf Stream, El Niño, weather patterns, and temperature. It gives you a very good overview of the chronology of climate change science, politics, and debate. It provides little in the way of solutions; however, it is well written, interlaced with personal, historical, and political anecdotes.

- » ***You Matter More Than You Think*** **by Karen O'Brien (Change Press):** O'Brien is an award-winning Professor at the University of Oslo, Norway, and an expert in adaptation to climate change. While people commonly talk about tipping points or disruption in business, economics, and technology, O'Brien shows that an equally momentous shift in the ways people think about climate and their places in the world is underway. When a critical number of people shift their thinking, a "quantum shift" can happen in the thinking of their whole societies.

Fiction and fairytales

Global warming makes a good story, tempting writers to use it as a basis for fiction. Like with movies (see the section, "Focusing Science on the Red Carpet," earlier in this chapter), however, the underlying science isn't always presented properly. One great book is *A Scientific Romance* by Ronald Wright (Picador). Inspired in part by H. G. Wells's classic *The Time Machine*, Wright's hero travels 500 years into the future, to a Britain transformed into a depopulated tropical jungle thanks to global warming. The author's beautiful prose and deft description make the situation seem all too plausible.

WARNING

One novel we don't recommend for satisfying your climate change curiosity is *State of Fear* (HarperCollins), a 2004 thriller by Michael Crichton. The book depicts global warming as a conspiracy concocted by conniving environmentalists. The story could be amusing if some people hadn't taken the book seriously. Crichton even attacked real scientists through his novel, with one character claiming that a prediction made by eminent U.S. scientist Dr. James Hansen 1988 about rising temperatures was off by 300 percent. Hansen himself has refuted that claim, showing that his projection was in fact remarkably accurate — an "inconvenient truth" for Crichton. (Check out Chapter 19 for more about Dr. James Hansen.)

Chapter **17**

Figuring Out How to Change before Global Warming Is Unstoppable

The big question hanging over humanity is whether there is enough time left before climate change becomes irreversible. If so, how quickly and in what areas must changes be made? Or will society continue to delay until nothing can be done to restore the relatively stable climate humans known for all of our existence? This chapter answers some questions about what changes are possible and how long those changes might take.

Asking Tough Questions

Fossil fuels have been a great gift from Earth to people — they have provided energy, abundant, easy to transport, easy to use, and flexible enough to meet all sorts of needs at relatively low (money) cost. Humanity would be nowhere near as well off if not for these ancient plant relations.

Today, though, the previously hidden costs of burning fossil fuels are evident all over the Earth. The air and the oceans are warmer; the ocean is getting more acidic; monsoons, hurricanes, blizzards, and droughts are violent and unpredictable; wildfires are more prevalent; and habitat for animals, plants, aquatic life, and insects (and humans' animals and crops) is changing. People are already on the move from where they used to live to where other people already live. The environment is becoming less and less optimal for human needs, and it's getting worse. So it has to change. The only possible way to change it is to stop burning fossil fuels.

Humans can choose to manage, or can choose to be overwhelmed by, changes to the Earth's climate. Here are the four big, really big, questions that need to be answered to get Earth back on a path suitable for humans (and for a lot of other living beings as well):

>> What needs to change to give humanity a reasonable chance to ensure a livable Earth?

>> Are these changes possible, and can they take place in time to prevent irreversible climate change?

>> Who needs to act to get the changes done? Who is standing in the way?

>> What can you do to get these actors moving as quickly as they must?

Making the Changes in Limited Time

Two hundred countries made commitments in the 2015 Paris Agreement and reinforced them in Glasgow, Scotland in 2021. Those agreements and the relevant technical reports call for three critical targets:

>> **Staying within the Earth's carbon budget:** From this time forward, no more than 400 billion tons of carbon dioxide can be allowed to enter the atmosphere.

>> **Net zero by 2050:** Making the world carbon neutral, so that by the year 2050 the amount of carbon emitted is balanced by the amount of carbon absorbed.

>> **"1.5 to stay alive":** To avoid the worst effects of climate change, the world's average temperature must not be allowed to increase by more than 2 degrees Celsius (3.6 degrees Fahrenheit) above pre-industrial levels (before the year 1850), and if at all possible, the average temperature should be limited to 1.5 degrees Celsius (2.7 degrees Fahrenheit).

All three targets are inter-related — each one affects the others. To meet them, there will have to be massive changes to the world's systems for energy production and use. Can the changes be made at all? Does the world have enough time?

We discuss in the rest of this chapter that there is a good chance that these targets can be met. Disaster isn't inevitable. The science is clear, the technology is available, and the benefits to humankind and the rest of the biosphere are obvious. The question is whether or not the world leaders have enough political and commercial will and determination to make the necessary choices. Humanity doesn't have to resign itself to some decreed fate. People, companies, governments: all have choices.

But these are big changes, with big questions attached:

» Can the changes solve the climate emergency?

» Are the changes physically and technically possible?

» If the changes are, can they happen fast enough?

» Are the changes possible without crashing the world's economies?

» Are the changes socially possible without massive shifts in your way of life?

Here you can judge for yourself if there are answers that can give humanity grounds for reasonably confident hope. And you see how your actions, combined with millions of others, can make those changes happen.

Staying within the carbon budget: Is it possible?

Of course, the world is still burning carbon fuels. The *carbon budget* is how much more carbon humans can pump into the atmosphere and still stay within those limits. The scientific consensus in 2020 was that, for a two-thirds chance to keep warming below 1.5 degrees, the remaining global carbon budget was 400 billion tonnes of carbon dioxide. That is, to stay inside the limits, no more than 400 billion new tonnes of carbon dioxide can be allowed to enter the atmosphere from here on out. That would mean that more than 60 percent of the world's known reserves of oil and gas and 90 percent of the reserves of coal will have to stay in the ground.

Humanity is now increasing the carbon dioxide in the atmosphere by around 36 billion tonnes per year. Divide the total of 400 billion tonnes overall allowed, by 36 billion tonnes per year, and you get 11 years of "budget" at current rates, starting in 2020. That's nine years from when we wrote this book in 2022.

WARNING

To repeat: If humans keep burning fossil fuels at the current rates, the world has nine years left before it gets to irreversible climate change. If you have kids, you know how short a time nine years is — just a moment, it seems, as children grow up. When you think of the long timescales for the Earth or the relatively short period of human existence, it's the blink of an eye.

Getting to net zero — Is it possible?

The IPC, the International Energy Agency (IEA), the International Renewable Energy Agency (IRENA), and other authorities have laid out scenarios in which humanity's consumption of fossil fuels is reduced by substituting renewable energy sources, and the Earth's capacity to absorb carbon dioxide is increased by changes in agriculture and forestry, so that by 2050 the carbon dioxide emitted is balanced by the carbon dioxide taken up. Most of these scenarios include human activities to take up a small fraction of the emissions. All include drastic reductions in fossil fuel use and increased efficiencies in power generation and use. But they all show that, with dramatic action, net zero by 2050 can be a feasible outcome.

Staying below "1.5 to stay alive" — Is it possible?

The IPCC reports from 2018 and later are firm: If the rise in average global temperature above pre-industrial levels (before 1850) isn't held to 1.5 degrees Celsius (2.7 degrees Fahrenheit), then there is a high likelihood of grievous damage from global warming. If global average temperature is allowed to rise above 2 degrees Celsius (3.6 degrees Fahrenheit), then there is an increased likelihood that global warming could keep on going and never be reversed, with catastrophic consequences for human civilization and other life on planet Earth. That's why people from the Pacific islands, whose entire countries will disappear if ocean levels are allowed to keep rising, have made the phrase "1.5 to stay alive" a slogan to keep in mind.

REMEMBER

This goal is tougher goal than net zero by 2050, because it has to happen sooner. To stay below the 1.5 degree limit, the IPCC calculates that humanity must reduce its emissions of carbon dioxide by 45 percent by 2030. It can be done, but this is where the going gets tough. So, if humanity is tough enough, it's past time to get going.

Addressing What Needs to be Done

The most immediate and urgent action to limit ongoing temperature increases is to reduce and eventually to stop injecting GHGs into the atmosphere. The most effective way to do that is to stop converting fossil carbon to atmospheric gases. But of course fossil fuels are the main energy source for the world's economy, are a major input into industrial processes, and provide energy and fertilizers for agriculture.

The following sections look at answering the same burning (well, hopefully not yet actually *burning*) questions to figure out what needs to happen:

>> Can the essential energy and raw materials provided by fossil fuels be replaced?

>> Is the replacement possible in time to keep global warming below 1.5 degrees Celsius?

>> Can the replacement be done without bringing the world's economy to a crashing halt?

Forecasting a path to 1.5 degrees

The International Renewable Energy Agency (IRENA) has provided forecasts for a pathway to 1.5 degrees. IRENA projects that fossil fuels, which provided 64 percent of the world's final energy consumption in 2018, will decrease to about 10 percent by 2050. At the same time, renewables, which provided 25 percent of electricity in 2018, will grow to 90 percent in 2050.

These projections from IRENA show one possible scenario. There are many others with different mixes of technologies. But they all show a very large increase in energy supplied from solar and wind. Remember those critical targets: a carbon budget of 400 billion tonnes, a 45 percent reduction in emissions by 2030, net-zero by 2050. What is required to get there from here? Can this happen to our worldwide energy systems now dominated by fossil fuels?

Electricity generated by the sun and wind has been growing at amazing rates. Electrical energy generated from the sun has been increasing by 44 percent per year for the last 30 years; electrical energy from wind has increased 21 percent per year over the same period.

UNDERSTANDING EXPONENTIAL GROWTH

It's not easy to imagine how small percentage changes per year can make things change so dramatically with time, but after you get it, you'll never forget, and you'll see it everywhere. When you read that "energy generated from the sun has been increasing by 44 percent per year for the last 30 years," that means that there is now 56,000 times as much solar energy available as there was back then. And that will double in a year and a half, so there will be 112,000 times as much. And 224,000 times as much a year and a half after that. Ten years from now, there will be about 4.5 million times as much — more solar energy than all of today's fossil fuel energy around the world.

This kind of growth (or shrinking) is called *exponential growth* — when a constant rate of change is the exponent. Exponential growth is very powerful. Consider this trick (you may try making bets on this one; it's a pretty sure winner with most folks):

> Say "I'll give you a million dollars today, or I'll give you a penny today, two cents tomorrow, four cents the next day, and so on for a month. Which do you choose?"

The answer (you probably got it right) is to choose the penny today that doubles every day for a month. After 30 days and 29 doublings, you'll have 5.4 million dollars, way better than the million dollars up front. If you get a 31-day month, you'll have 10.8 million. That's the power of exponential growth. And that's what's happening with the growth of renewable energy.

Along with numerous other authorities, the International Energy Agency (IEA) believes that the technologies needed to achieve the necessary deep cuts in global emissions by 2030 already exist, and that these rates of growth will make for "a complete transformation of how we produce, transport and consume energy." The IEA projects that by 2050, 90 percent of global electricity generation will come from renewable sources and 70 percent of that will come from solar and wind. The technology available now is enough to get this done.

But that's just the beginning. Two major factors will drive this growth even more rapidly:

>> Both solar and wind generation keeps getting more effective — more usable energy per hour of sunlight or wind.

>> As the technology develops, the cost keeps going down with no end in sight.

The combination of the need to replace fossil fuels, the increasing efficiency of renewables, and their continuously decreasing cost means that fossil fuels are likely to be simply uncompetitive in most markets in a decade or so. This is when disruption happens. It's inevitable. The only questions are about how long it will take.

Understanding Disruption

Disruption is the replacement of entire systems by newer, cheaper, safer, and more accessible ways of doing the same things, and opening up of entirely new opportunities impossible with the old ways. Business disruption often means that the existing players are unable to adapt, they no longer have strategic choices to help them survive, and they go out of business.

These sections look at some familiar examples of earlier disruptions. You get a look at the (predictable) reactions of the existing industries and their supportive governments and financiers. And we examine the amazing speed and reach that the coming truly revolutionary disruption will have — perhaps even enough to save the planet.

Predicting what disruption of fossil fuels looks like

The choice of the fossil fuel industry is so far clear: Deny, delay, drill, dig, and burn. They keep making money as long as it's allowed and as long as their product is attractive to customers. Refer to the section, "Ending the dominance of fossil fuels as humanity's main source of energy," later in this chapter to see what other choices oil executives have. More and more evidence shows that, in the biggest markets, coal, oil and methane might not be competitive for much longer. To be truly disruptive, the energy produced by solar and wind generators must displace most of the energy produced from fossil fuels. The renewable energy will have to be better, cheaper, more easily processed, distributed and used, and more efficient in existing applications that use fossils. It should also open entirely new applications and possibilities. Is this possible?

Renewable sources of power are coming on at a gallop. In many cases they're already cheaper and easier to process, transport, and use. They've always had the technical potential to disrupt the fossil fuel business; now they have the business and market momentum as well. That momentum is building both greater capacity and reduced costs. More and more energy systems like vehicles, buildings, and manufacturing systems will be electrified. Fossil producers will be able to sell only the cheapest and sweetest products, like Saudi oil (produced at around $2.80 per barrel from the oil deposits underneath what is planned to be the world's largest solar generator — a double win). Products like oil from the tar sands or from fracking will simply be too expensive to sell in most markets, and most current producers will go out of business. That's disruption.

Recognizing the cost of disruption

Disruption isn't free for people and companies in the businesses that go out of fashion. As we discuss in the section, "Staying within the carbon budget: Is it possible?" earlier in this chapter, to stay within the carbon budget about 60 percent of the oil and gas assets and 90 percent of the coal have to stay in the ground. So really it will have almost no market value at all. In the case of the fossil fuel companies, estimates suggest the loss of the current value will be somewhere between $11 to $15 trillion, which is a lot of money, a lot of shareholder value, a lot of defaulted debt.

As we discuss in the following sections, that's why many savvy institutional investors are getting out of fossil fuels (check with the management of your retirement fund, if you have one), more than half of the world's banks are reducing their exposure (at least, they're not lending any *more* money), and the larger oil companies are selling assets to smaller hungrier companies who think they can move faster to develop and sell the products before the door closes forever. It's a tough time to be in the oil and gas business and a horrible time to be in coal. And it's a dangerous time for financial markets, as all that value in the fossil fuel assets no longer goes up in smoke.

The reaction of the exiting industry players is predictable and understandable. They're in a tough spot. Consider, for example, the reaction of executives in four affected sectors:

>> Fossil fuels producers and electrical utilities

>> Bankers and investors

>> Industrial energy consumers

>> Political leaders

The problem for a fossil fuel company

Imagine yourself to be an executive in a fossil fuel company. You can see the writing on the wall — you know that your products aren't healthy for the world, and you know that eventually you'll have to stop selling them. Today you have shareholders, bankers, employees, and customers. They all depend on you to keep doing what you're doing, otherwise they lose. If you say, "We're getting out of the fossil fuel business," what do you think your bankers will say?

Those are the same bankers who lent you millions of dollars because you promised them a good rate of return, based on the value of your oil and gas assets. You can't just abandon those assets — your bankers will foreclose on your business. You can't change your infrastructure of wells, mines, processing plants, and pipelines into technology to make energy from renewable sources. What are you going to do with all that stuff? What about your employees and suppliers? They work for your company for money and security and a sense of a good job well done. What else have you got to offer? None of your business colleagues want to buy yours; they're in the same boat as you are. And your customers want oil and gas and derivative products — if they wanted electricity, they'd buy it. What do you do? You try to keep your shareholders happy by making as much as you can as long as you can. You resist all the efforts and forces, telling you to shut it down because you're killing the planet. You have to do your job. Right up until the end, you resist with all your strength.

The problem for a power utility

Speaking of electricity, what do you do if you're on the board of a big power utility? Your business has been built the old way — a big central plant burning coal, making electricity and tied to a grid of wires to deliver it to industrial and household customers. It cost millions, maybe billions, of dollars to build that big plant and that distribution grid. So your friendly local government, wanting to encourage business development and improved lives for voters granted you a monopoly on selling power in your area so you could have predictable revenue for years to come, which enabled you to borrow all that money. And it has worked, for years and years. Your customers had a dependable supply of power, you had excellent revenues and profits, and all was good. You didn't need to worry too much about paying back all that borrowed money because after all your revenue was pretty much guaranteed.

Suddenly, without much warning, your customers want to put solar panels on their roofs and windfarms in their fields and sell you power, cheaper power than you can make, and they want to use your grid to distribute it to others. What do you do? You resist, of course — you can see that your ability to pay off your debt isn't so secure any more, you can see that your utility will never built another big plant of the kind you're so familiar with, and you see in your dark moments the time when your generating business is over, and you're just a power distributor for thousands of renewable sources. What do you do? You try to keep your bondholders and bankers happy by making as much as you can as long as you can, and you resist all the efforts and forces telling you to change because your product is no longer competitive. You have to do your job. Right up until the end, you resist with all your strength.

The problem for a banker

If you're a small player and your investment is easy to cash out, you flee as millions are doing around the world. But what if you're a big lender, and your loan is backed by the collateral of those fossil fuel assets or that coal or nuclear-generating station? You're stuck. You might want to get those loans back, but how? The assets are worth less and less each day, you've still got tons of money at risk, and you can't possibly get paid back unless the company or the utility continues to operate and make profits.

The problem for an industrial energy user

If you run a shipping company, you need the cheapest fuel to get your cargoes around the world. Maybe you're concerned about climate changes — you can experiment with fuel cells, batteries, solar and wind power, and hull shapes for greater efficiencies, but really you're pretty much stuck with heavy fuel oil. You're not an energy provider. You need someone else to provide you with a new fuel that reduces the pollution and greenhouse gas (GHG) emissions. Run an airline? Much the same. Trucking company? Lots of promises, no electrical long-haul over-the-road trucks yet. Mining? Moving construction materials, steel, wood? Still waiting for electric shovels, loaders, and so on? The list goes on and on. Even where people are ready and willing to shift their energy burdens away from fossil fuels to renewables, the necessary bits and pieces just aren't ready yet.

What if you make steel or concrete? You use a ton of energy, and you put out more tonnes of GHGs. Your production facilities are big, not easy to replace or upgrade even you had alternatives. You know about new means to make concrete with lower carbon dioxide emissions, but they won't work in your plant. You know that mini-mills are using electricity to recycle steel, but your big outfit relies on coal to make its heat.

So you carry on — you can't wait when it comes time to replace equipment, so you buy another gas- or diesel- or heavy-fuel-oil powered machine. Even though you want to make the problem better, the actions you have to take are bound to make it worse. You're perfectly ready to have your suppliers disrupted and to put new technologies to work, but your choices are few to none.

Considering other benefits to disruption

This massive world-around change in energy systems will have other beneficial effects. Developing countries will be able to leapfrog right over old-fashioned fossil fuel systems, to build their new economies on locally generated and less expensive energy. No more need for billions of their scarce dollars to flow to the richest of the rich in the industrialized countries. India and China, both much too dependent now on coal, are enormously well-positioned to develop in this manner.

DISRUPTION IS NOTHING NEW

You can find numerous examples of whole industries and ways of life being disrupted with amazing speed. Horses by automobiles, film cameras by digital cameras, landlines by smartphones — each within 10 to 15 years. In each case the existing industries were firm: "the new technologies would take a very long time to develop," "the existing way of doing things would continue for years," "the changes would cost enormous amounts of money and thousands of much-needed jobs," and "the change should be resisted rather than encouraged."

Still, those old technologies that seemed to be cornerstones of a way of life are done and gone. And the technologies that replaced them created such massive opportunities for new activity that society is far better off. Of course, they brought with them new problems as well — car accidents, intrusive snapshots, screen addictions — but they mainly brought benefits.

Every time the existing players resist, their supporters in finance and government help them resist. Every time the existing players get displaced, their supporters move on to become cheerleaders for the next thing. And every time, the disruption that people feared would be so costly leads to massive new opportunities and increases in social well-being. Except, of course, for those disrupted.

Those about to be disrupted complain about the costs. Of course they do — so would you if you were in their shoes. But the overall benefits far outweigh the particular costs. The Institute for New Economic Thinking at Oxford University says:

> "Concerns about costs of rapidly decarbonizing the global energy system have been a barrier to implementation, but . . . a greener, healthier, and safer global energy system is also likely to be cheaper. We find that, compared to continuing with a fossil-fuel–based system, a rapid green energy transition will likely result in overall net savings of many trillions of dollars."

It's mathematically possible to build out solar, wind, and battery capacity rapidly enough that they could be the world's major energy suppliers, displacing fossil fuels, quickly enough to meet the goals of a 45 percent reduction in GHGs by 2030 and net zero by 2050.

But mathematically possible doesn't mean that it's certain. As ReThinkX says, and as the IEA confirms "we already have the technologies to achieve net zero emissions." But more than likely the current rate of installation of these new technologies, although very rapid indeed, isn't quick enough to meet humanity's goals. What can be done to speed it up?

Ending the Dominance of Fossil Fuels as Humanity's Main Source of Energy

Saying that the replacement of fossil fuels is bound to happen is easy. Renewable will just plain outcompete them. But is it bound to happen fast enough to meet the three targets of a limited carbon budget, a limit on global temperature increase, and getting to net zero by 2050? These sections examine fossil fuel's tight grip on humanity.

Changing the ways humans use energy and power to get things done

A terrific amount of innovation and invention is happening right now. In response to the climate emergency, humans all over the world are coming up with solutions. We describe some of them in Chapter 6, at the personal level, the micro-scale of individuals and households. Solar panels on roofs are becoming commonplace. Chapter 14 shows that many industries are working at the medium scale within their facilities, plants, ships, and planes to reduce emissions and increase efficiencies. Local governments are changing the ways their towns and cities operate to reduce individual trips in individual vehicles, to encourage better home insulation and replacement of fossil fuel–powered furnaces and appliances with more efficient and zero-emission electrical versions.

Action at the large scale is also happening, like the following:

» New solar photovoltaic (PV) plants are being built in places far and wide — from Arizona to Alberta to Abu Dhabi — and some are even planned for Saudi Arabia. Thousands of megawatts (as big as all but the very biggest hydro dams) of solar energy capture, resulting in available power at prices less than a tenth of the most efficient hydro systems.

» Windfarms on open land and on the ocean are being built, wired into international grids for backup and load sharing.

» Green hydrogen (not hydrogen made from natural gas) and ammonia fuels are rapidly being developed, using renewable energy sources to make clean fuels.

» Batteries are continually being improved for higher power density, longer service life, easier recycling, and with fewer hard-to-find materials.

Changing what political leaders see the best interests of humanity

Although all of the actions in the previous section are having an effect, they can't do enough on their own to halt the rise in global temperature. It's a little bit too little, a little bit too late. The projections made by the best in the world — the IEA, the International Renewable Energy Association, NASA, the Intergovernmental Panel on Climate Change (IPCC) in its latest reports, and the discussions at the United Nations' recent COP26 in Glasgow — tell you that today's movement isn't going to be fast enough to solve the problem. Acceleration is essential, or the world won't be able to keep to the given limit of 1.5 degrees Celsius warming above pre-industrial temperatures.

Figure 17-1 shows the problem. Politicians and industry keep on trying to convince humanity that the important goal is net zero by 2050. This draws attention away from the first and toughest problems of reducing emissions by 45 percent before 2030 and of staying within the overall carbon budget. If these two goals aren't met, the carbon budget will be way overspent, and net zero just won't matter. So, the whole program of decarbonization has to be accelerated. You can see in the figure how it works: To get to a 45 percent reduction by 2030 and to stay within the budget, steep emissions reductions have to start now. But if the only goal that's reinforced is net zero by 2050, the major reductions can begin much later. That's why those who are resisting the disruption don't talk much about the carbon budget or the required 45 percent emission reductions by 2030 — they much prefer to keep you focused on net zero by 2050.

REMEMBER

That acceleration has to come from governments. Only governments and international institutions can create the legal, regulatory, and financial environments to overcome the resistance to change and to encourage the deployment of renewable energy sources even more rapidly than is happening now. Governments at all levels — local, regional, national, and global — need to lead this charge.

Like the incumbent energy industries and energy users, governments are embedded in existing systems. The fossil fuel business is everywhere, with tremendous money, influence, and power. In democracies, many politicians depend on donations from individuals and corporations for political success. And the individuals and corporations who have the money are more often willing to donate it to individuals who will advance and protect their interests; if a fossil fuel consortium funds a politician, you can expect that support for that industry will come high on that person's agenda. In some totalitarian states, governments are largely dependent on the revenue from exporting fossil fuels. You shouldn't expect those governments to move very fast to put themselves out of business. And the fossil fuels are still the most readily available energy sources for developing countries as well.

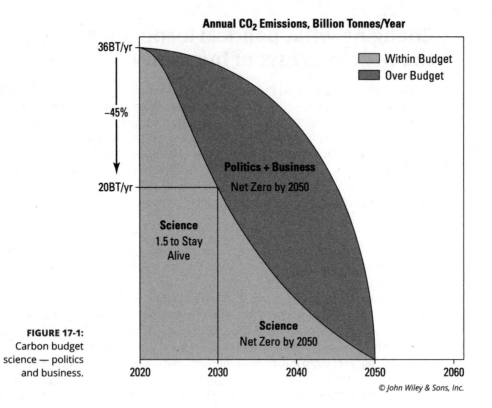

Annual CO$_2$ Emissions, Billion Tonnes/Year

- Within Budget
- Over Budget

36BT/yr

−45%

20BT/yr

Politics + Business
Net Zero by 2050

Science
1.5 to Stay
Alive

Science
Net Zero by 2050

2020   2030   2040   2050   2060

© John Wiley & Sons, Inc.

FIGURE 17-1:
Carbon budget
science — politics
and business.

So then, what can get this all going more rapidly? What can get government and international leaders to take action that won't please influential supporters? What action could those leaders take anyway, even if they wanted to?

The best we know is fast action to reduce GHG emissions might keep earth below the 1.5 degree limit. But the necessary changes in the following *all* must happen in a short period of time. Each is necessary, none on its own is sufficient:

>> Energy production and power distribution

>> Energy use in industrial, commercial and household systems

>> Financial systems

Understanding How Disruption Can Be Accelerated

The necessary actions to accelerate the disruption aren't hard to identify. The issue isn't *what* citizens must push governments to do, the issue is *how* to push them to do it.

The IEA says:

> "To get renewables on track with net zero by 2050, governments not only need to address current policy and implementation challenges but also increase ambition for all renewable energy uses. . ... Governments should also consider targeting much more economic recovery spending on renewables while also putting in place policies and regulations enabling higher mobilization of private capital."

Can you expect governments to lead like this?

The means aren't hard to outline. For starters:

>> Stop issuing licenses to grow fossil fuel production.

>> Shift subsidies now paid to fossil fuel producers to developers of renewables.

>> Shift support for new fossil infrastructure like pipelines, liquified natural gas (LNG) plants, and shipping terminals to support for grids.

>> Stop wasting money on expensive new hydro or nuclear plants.

>> Provide simple low-cost financing for householders and business to

- Electrify vehicle fleets

- Improve structures, appliances, and machinery

- Put solar PV on the rooftops

>> Put more research money into green hydrogen and ammonia as alternate fuels.

>> Change buying practices to favor renewable energy choices.

>> Get utilities to move rapidly on becoming market makers for distributed suppliers of energy.

There won't be much need for all the various carbon pricing solutions (*carbon pricing* means deliberate regulations, carbon trading schemes, or direct taxes that add to the final cost of fossil fuels by adding to the industry's costs and profits some portion of the real costs to the planet, in addition to the industry's costs) that we describe in Chapter 10. Nobody will have to add costs to carbon when it's already priced out of the market. Likewise, nobody will need to be restricted in their choices for personal energy use — the disruptions will make energy cheap and plentiful. Humanity will certainly need to pay more attention to overconsumption — it would be tragic if the disruption prevented burning the world up only to encourage using it up.

REMEMBER

What governments need to do is to get behind the disruption, instead of standing in front pushing back against it. An observer at COP26 made the comment that "governments are standing on the fire hose — all they need to do is to let it flow."

Most of the readers of this book live in democratic countries. And in democracies, politicians mostly have to follow the crowd — after all, that's where the votes are. (That's not always true – there are *real* leaders out there. We describe some of them and their works Chapter 19.) But most elected politicians aren't paid to be brave. In addition to the technical and business disruption that's happening now, they need to see and feel that a *social movement* is happening as well. Then they'll be jumping over each other to take the lead. There's a famous quote from a French politician in the 1870s: "There go my people. I must find out where they are going so I can lead them."

That's where citizens, perhaps you, come in.

Getting There from Here: A Conversation among Citizens

Citizen involvement, coming initially from a small base, can grow very rapidly until a critical mass of people are involved. And after that critical mass is reached, the movement can become unstoppable. Already, more than 60 percent of the adults in the United States, and more in Canada and Europe, are alarmed or concerned by what they can already see as the climate changes. Humanity is now experiencing a 1.1 degree Celsius rise in global average temperature, and the effects are already pretty scary.

Fewer and fewer people still think climate change isn't happening, and pretty much everyone now knows that it's happening because of humans burning fossil fuels. The effects are visible all around — many people are personally affected, and it's increasingly clear to more and more people that the world can't go on as it has in the past. So more and more people care about climate change, and more and more people around the world are adding their voices to the conversation. Your voice counts too.

When such a big part of a population, a nation, or a society start to feel the same way, things change. As we discuss in the section, "Changing what political leaders see the best interests of humanity," earlier in this chapter, politicians are generally followers, not leaders. They're bound to follow when your voice is added to the voices of millions of others. You can tell them, in no uncertain terms, that you won't put up with their lack of action, or worse, their resistance to any action at all. Then and only then will they move to encourage the disruption rather than resist it.

And it won't take much encouragement to make this happen.

Authors Katherine Hayhoe and Karen O'Brien in their books encourage you, and everyone around the world, to have conversations with each other about climate change. Such conversations will almost inevitably lead to more common ground than to opposition, when people see how each one is affected by the ongoing climate emergency and how each person sill benefit from the massive disruption to occur. After a critical mass of individuals is in the practice of speaking to each other, then the social will for change will become dominant. And that will force politicians to take action to respect the wishes of their constituents.

REMEMBER

Social disruption is generally seen as a negative thing — a term used to describe the alteration, dysfunction, or breakdown of social life. But it can also be positive — think of the movements for huge social change with which you're familiar like those to end slavery or to emancipate women. Here you are, here we are all the people in the world, faced with a common problem — the current industrial system based on fossil fuels is endangering life and today's diverse civilization. If ever there was a time for a mass movement, that time must be now.

It takes surprisingly few people in a population to make changes that reverberate throughout. A recent study published in the magazine *Science* concluded that 25 percent of a population is the tipping point for overall change. That is, if 25 percent, one person in four, believes in and communicates a feeling, then that feeling can quickly become dominant in the overall population. Conversation, ordinary chats with colleagues, neighbors, friends, and families, moves the ideas around. It doesn't require a megaphone, or a social media platform, or a commanding perch — just day-to-day conversation about something of vital concern to everyone.

Humanity is already past the 25 percent who care.

Almost everyone agrees that if things can be made better for our children, we should do all we can to make it so. Almost everyone agrees that our environment is precious, and that it should be protected for us, our descendants, and the other beings on the Earth. These aren't matters in dispute.

The loudest voices so far have been those whose economic interests are threatened. But those voices won't be loudest for much longer — even as they can do their very best to delay things, they know that the coming disruption is inevitable.

The logic of technical and business disruption applies here as well: Within a short time, the "new" ideas will be seen to be the common wisdom. The politicians will recognize that their livelihood and their positions of responsibility won't survive unless they recognize the social disruption happening around them. Then, and only then, will they take the necessary actions so obviously required.

Putting the climate emergency behind us will be possible if, and only if, both kinds of disruption happen now. Then the world can move to meet the two critical targets of a 45 percent reduction in emissions by 2030 and net zero emissions by 2050. Technical and business disruption can do the job. Social disruption, your voice and those of others, is needed to pull and push it along.

Yes. It can be done. And your children and grandchildren will look back on this time with gratitude and pleasure.

6

The Part of Tens

Identify the most important things you can do to affect the climate emergency.

Meet a line-up of inspiring individuals who are playing a major role in applying solutions.

Debunk common climate change misunderstandings and myths.

Discover some online resources to keep you informed — so you can do some myth-busting on your own.

Chapter **18**

Ten (Plus One) Things You Can Do to Fight Climate Change

We don't blame you if you flipped ahead to this chapter before reading all, or any, of the rest of the book. In fact, we congratulate you — you want to do something about climate change. What you can do to help slow global warming depends on where you live, the resources you have, and how much time you can give. You may not be able to slap solar panels on your roof tomorrow, and you likely can't ditch your car for an electric vehicle by next Tuesday. But you can make simple changes that have a big impact. This chapter offers some solutions that you can implement right away.

REMEMBER

You may be skeptical that what you do can have any meaningful impact. Rightly so because fossil fuel companies and their allies have spent a ton of effort and even more tons of money trying to convince regular people that climate change is all their fault. They say they're not responsible and that they're just responding to consumer demand. However, smart people, like you, are seeing though that deception: It's common knowledge now that the fossil lobby has known for decades that carbon emissions have been changing the atmosphere, and that they knew about the threat that poses to the climate. They've been paying the same consultants that used to work for the tobacco companies to run the same strategies of denial and to insist that the responsibility is with the consumer, not the seller. If you change your ways, they say they will too. And if you can't change your ways, well then they say they can't possibly change theirs.

The companies are trying to convince you that your personal actions have gotten humanity into this fix. Sure, each of us has a role — both as citizens and as consumers of energy and goods. But the big problems are in the big industrial, economic, and political systems. Climate change is not *your* fault.

You can make simple changes that have a big impact. This chapter offers some solutions that you can implement right away. For more on how your personal choices impact the climate, check out Chapter 6.

TIP

The biggest opportunities you have to make changes as an individual only come up over years — purchasing a new car or truck, buying a new heat pump to replace your gas furnace, getting that new fridge or stove, renovating your home, putting solar panels on your newly insulated roof, and so on. They're not day-to-day purchases. When you're in a position to make these big decisions, you have some cool (pun intended) options. These options can have large impacts on the too-warm planet. Your choices now can give you much better service and reduced costs, to your benefit as a conscious consumer, along with decreased carbon emissions and reduced energy use, to the world's benefit as a conscious citizen.

Greening Your Car

The *life-cycle cost* (the cost to purchase, operate, and maintain something over its lifespan) of a new electric vehicle (EV) is already less than an equivalent car powered by an internal combustion engine.

The purchase price is coming down fast. And maintenance is a breeze: 20 parts in an electric motor, thousands in a combustion engine. No radiators and hoses and clamps and leaks. No transmission in most EVs. Brakes that are only used to keep the car stopped, not to slow it down. Range between charges in the hundreds of miles. And, contrary to some of the propaganda you may have heard, the environmental costs of making an EV, including the batteries, are less that the environmental costs of making a conventional car.

Even if you had to use conventional financing and if no tax credits were available, if you could afford an EV now for your next purchase, you'd be better off. Government tax credits and financing alternatives are coming on stream in many countries. By the time you come to your next vehicle purchase, the choice will be increasingly clear. And that's not even knowing that you're not making a real contribution, but you're also setting an example for others.

Upgrading Major Appliances

"Now you're cooking with gas" was an advertising slogan cooked up to sell more methane (the phrase "natural gas" was itself a way to rebrand the unattractive sounding "methane"). But now cooking with electricity is the way to go. *Consumer Reports* says that smooth-top or induction electric stoves and ranges now outperform gas equivalents in nearly every way. Even electric barbeques are coming to rival gas versions. So that new stove? Make it electric.

TIP

Other new major appliances are already electric. Manufacturers and retailers recognize that energy is now a factor in how you're going to make your buying choices, so they're competing on energy efficiency as well as features, convenience, and maintenance. Start the habit of looking for energy-efficiency labels, such as Energy Star. That next big appliance you buy will be much better for the planet.

Buttoning Up Your House

The key to creating an energy-efficient house (whether you want to keep the heat in or out) is insulation, insulation, insulation. Heating and cooling costs make up a whopping 80 percent of your energy bill, so seal up your doorways and windows, and make sure that you have the proper insulation for your crawl space, attic, and walls. Some of these changes like sealing doors and windows are simple fixes with immediate payback and can be done without much money or time. Others, like adding full-scale insulation, upgrading windows, or installing solar panels on your roof, can take significant time, effort, and money, but they last for decades and save you significantly in energy costs, all while cutting back on greenhouse gas (GHG) emissions.

REMEMBER

Start with an energy audit so that you know where your dollars can give you the best results.

Making Your Daily Living More Energy Efficient

Your car, your house, your work, and your day-to-day and occasional activities all affect your personal emissions. Your choices are important here as well. Here we discuss a few things you can do at home and when you're driving.

At home

Be conservative with your purchasing choices. When you buy electronics and appliances, only buy what you need. If you get a bunch of gadgets that you're lured into purchasing, you end up consuming more electricity than you intend. And you end up making more waste because more than likely those tech gadgets will be outdated in a year or so and then you'll have all those wires and gadgets to dispose of.

TIP

The energy habits that you have in your daily life make a difference, as well. Here are a few energy-saving changes that you can make to your daily routine:

» Use a drying rack or a clothesline rather than the clothes dryer. If you do use the dryer, fill it up and don't run it longer than necessary.

» Switch to compact fluorescent light bulbs or light-emitting diode (LED) bulbs, and turn on the lights only when you need them.

» Shut down your computer when you're not using it.

» Buy products that are reusable and recyclable or that include recycled content.

» Choose energy-efficient forms of travel, such as the train or bus, for short-haul trips. To get to work, take your bike, the bus, or the subway. Avoid unnecessary trips altogether.

On the road

Fuel emissions from transportation account for about 16 percent of global GHG emissions (or 24 percent of emissions from energy use, not including deforestation). Transportation accounts for almost 30 percent of energy use in the United States, and more than half of that is from personal vehicles. Most of the rest of the world uses considerably less, but it's still an important factor everywhere.

As painful as it sounds, you can best cut down your car emissions by not owning a car at all (although without significant changes in land-use planning and access to mass transportation, this will be more difficult for some than others!). Even if you're fortunate enough to have an EV, the Earth is still better off if the electrical energy is saved whenever it can be.

REMEMBER

If your car has an internal combustion engine (ICE), you can still drive smart. Efficient driving is climate-friendly driving. A lot of little changes can substantially cut your car's emissions:

- » Drive as smoothly as you can. Avoid rapid acceleration and sudden braking.

- » Carpool when you can.

- » Turn off your car when you stop for ten or more seconds. This applies when you're pulled over to stop, sitting in a drive-thru, or delayed by traffic construction that brings everyone to a standstill — not if you're in the middle of traffic at a stoplight. In the 1950s, you did use more energy re-starting your car a lot, but that's not true anymore.

- » Do all your errands at the same time, instead of spreading them throughout the day.

- » Keep your car's engine clean, up-to-date, and running efficiently by taking your car in for regular checkups (or doing it yourself).

- » Fill your tires to their ideal pressure point (which you can find written right on the tire) to use less fuel.

- » If you have a ski rack mounted on the roof of your car, remove the rack in the summer to reduce drag, which helps reduce fuel use.

At work

Many people spend the majority of their waking hours at work. Your employer's building and transportation fleet (if your company has one) presents two major projects that are just waiting to be tackled. If you nose around, you may find that your colleagues also want to turn your office green. If your employer wants you to work from home now, as so many do in these days of pandemic illness, you're already being greener than the daily commute required you to be before.

TIP

But if you're still in an office or other centralized location, here are a few suggestions that you can take to your boss, building manager, or other decision-makers at work to get started:

- » Establish an energy policy that cuts down on the company's or organization's energy use and sets a target for reducing emissions. See www.theclimate group.com for a case study that may relate to your workplace.

- » Put a recycling and composting system in place.

- » Post signs or notices that remind people to turn off their computers, lights, and other office equipment when they leave.

- » Organize monthly lunch programs and have a speaker come in to talk about solutions to climate change.

Many of these tasks mean taking an issue to a project manager or a boss, or even to the director of maintenance or facilities management. Forming a committee or advisory team with your colleagues can make you more effective and successful — you can find strength and support in numbers.

REMEMBER

When a company stands for more than profits, employees show greater corporate loyalty and improved morale, which results in better job performance and a drop in absenteeism.

Going Vegetarian or Vegan (Sort of)

Eating fewer meat products can help reduce your carbon footprint. Cattle in feed-lots and pigs and chickens in intensive factories are hugely energy-intensive. Cattle pose another problem: Cattle and other ruminants like oxen, sheep, and goats (*ruminants* are mammals that partially digest food and then bring it back up to chew again — "chewing the cud") have a unique digestive process that allows them to convert otherwise indigestible grasses to food energy. In that process, their *rumen* (a specialized stomach) generates lot of methane (natural gas). The only way out for that gas is up the throat through the mouth to the atmosphere: Ruminants belch out almost 4.6 percent of global emissions of GHGs. Further-more, feed production (including land-use changes like deforestation to create grazing land) and processing for all those burpers puts out about 5.5 percent. So cutting back on meat can have a real effect.

TIP

You don't need to go full-out vegetarian. Try starting with going one or two days per week without meat. The next best solution is to try to buy locally raised or wild meat, which usually comes from smaller, less energy-intensive operations. Chapter 6 covers more options for reducing your carbon footprint through your food choices.

Reducing Food Waste

Agriculture contributes around 30 percent of the GHG emissions around the world. Much of that's in rich countries, where massive overuse of fuels, fertilizers, and pesticides supports production beyond the sustainable capability of the land. And, in rich countries, almost 30 percent of the food grown is simply wasted, thrown away because it's not perfect or stale-dated to satisfy marketing needs, not safety issues. Worldwide about 8 percent of total GHG emissions are simply from wasted food, according to the U.N. Food and Agriculture Organization (FAO).

In poorer countries, lots of food is lost each year to inadequate storage infrastructure that allows access by pests and diseases or molds. All over the world, production of staple crops is threatened by extreme heat, drought, floods, and uncertainty. Farmers are experienced at dealing with risk and variations in expectations for temperature and weather, but not with the sort of massive uncertainty brought by the climate emergency. What do you do if it just doesn't rain for years? What do you do if the crops you've relied on for decades are no longer productive in their new circumstances?

New, or in most cases, revived old practices like *zero-till* (seeding this year's crop directly into soil without plowing, discing, or harrowing after last year's harvest) or *regenerative agriculture* (farming and grazing practices that reverse climate change by rebuilding soil organic matter and restoring degraded soil) can help to deal with this problem to grow reliable crops, to heal and rebuild the soil, and to make agriculture a net contributor to the climate. After all, plants take carbon dioxide out of the atmosphere to grow and supply oxygen. It's not the basics of agriculture that create the emissions, it's energy-intensive modern farming that makes it so costly to the atmosphere.

TIP

Support small-scale local agriculture. Don't buy so many foods with a huge carbon content just from transportation — avocados from Mexico, for example — basically anything way out of season where you live has been brought to you by trucks and trains and airplanes. And use the food you buy or grow.

REMEMBER

Any choice you make can help reduce your emissions.

Supporting Clean, Renewable Energy

You can back the development of clean, renewable energy in a number of ways, depending on where you live.

TIP

Here are the two most common ways to make a significant impact:

>> **Make the energy.** You can make energy yourself by using options such as solar energy to produce hot water and generate electricity. Rooftop solar gets cheaper every day; if you have a roof available and you live in an area with adequate sunshine, investigate whether your local power utility or government will support your installation. Chapter 13 covers other options that you can use to make energy.

>> **Buy the energy.** You can purchase energy from a company that uses low-emission energy sources. Green power can be purchased in most provinces in Canada. Companies also cover 14 states in the United States, with similar companies across the U.K., and in every state in Australia (see Chapter 21 for website resources on green energy providers).

Being a Smart Investor and Encouraging Smart Disinvestment

The companies and utilities that produce and use fossil fuels are very sensitive to the choices made by the shareholders — the CEO doesn't need to care at all about global warming if their paycheck and stock options are worth less every day. So, if you're an active trader, investor, or long-term shareholder, you can have a direct impact on the companies with which you choose to get involved.

Millions of individuals are shedding their holdings in fossil-fuel companies or in utilities that are still using fossil energy to generate power. Tens of thousands more are getting engaged with or moving their investments to activist shareholder groups. Imagine how surprised the old-timers at Exxon were when activist investors installed three directors on its board to push the company to reduce its massive carbon footprint.

TIP

If you're not an investor yourself, you may be part of larger groups that are — bank, pension funds, and employer are the prime examples. You can write letters, make phone calls, contact anyone you can, and make it clear that you too are one of the growing number of people calling for the fund to disinvest and get out of fossil fuels and related investments. More than $15 trillion of investments have been pulled out of the fossil fuels as part of the worldwide disinvestment campaign. Your bank or credit union, your employer, your pension fund — they all react to your interests as well, given enough pressure. Pile it on. You can find much more about this sort of environmental, social, and governance (ESG) investing in *ESG Investing For Dummies* by Brendan Bradley (John Wiley & Sons, Inc.).

Getting (or Making) a Green Collar Job

Many jobs offer the opportunity to contribute directly to climate change solutions. And you're not limited to working within an environmental nongovernmental organization (NGO) — though we highly recommend this option! The following list gives you a glance at the kind of climate-friendly jobs that are out there today, for a range of qualifications, skills, and interests:

>> **Architecture and design:** The world needs green building construction and design, and this field is looking for architects and designers with vision. You can help make everything within a building — from materials to energy systems — green. (We talk about greener buildings in Chapter 14.)

>> **Building/contracting:** You can offer your clients Forest Stewardship Council certified wood (see Chapter 14), materials (such as plastic, stone, wood, insulation) made with recycled content, and environmentally sound materials (that are biodegradable or able to be recycled at the end of their lifecycle) to help them make informed choices and help reduce emissions adding to climate change.

>> **Education:** As a teacher at any level, you have many opportunities to integrate climate change issues and solutions into your curriculum. You can even join one of the many organizations of educators who work together to have environmental issues put into standardized public education.

>> **Engineering:** Engineers who want to work in climate-related fields are in high demand. The long list of fields includes environmental assessments, water resource management, carbon capture and storage, and GHG assessments.

>> **Government:** With more government programs starting up in various countries, new positions are created regularly. Look for posts within the departments that deal with climate change, the environment, energy, sustainable development, agriculture, and forestry. (Chapter 10 outlines some of the actions governments can take to help fight global warming.)

>> **Higher education:** Being a student isn't exactly a job (although you can find many climate-related research grants available), but it presents a way to take your knowledge about climate change to a higher level. Many universities and colleges around the world now offer programs that focus on energy, the environment, climate change, environmental management, climate law, and more.

>> **Music:** When you're a musician, you have people listening to you all the time (or, at least, that's your goal). You can heighten climate change awareness through your lyrics or by speaking to your audience between songs.

>> **Nongovernmental organizations (NGOs):** The campaigns and projects available through NGOs allow you to work directly on an issue and advocate for change. Whether you lead the campaign or assist in the office, your job contributes to the overall effort of effective projects. (Check out Chapter 15 for more about NGOs.)

>> **Visual art:** Being a visual artist gives you a unique opportunity to express anything and everything about climate change. Photos, film, and all kinds of visual works of art can effectively communicate the urgency of global warming.

>> **Writing/journalism:** If you write professionally, people read what you have to say. You have a wonderful opportunity to introduce your readers to climate change issues. (Chapter 16 explores how the media covers global warming.)

REMEMBER

You can find endless opportunities, whether you're an entrepreneur or a job seeker, if you go looking for a green collar job. Making a difference is a terrific feeling.

Helping To Make Change Where You Live

Still, no matter how much you change your personal purchasing and consumption choices, much larger system-wide changes are still needed. Your voice can be a vital part.

By launching a local campaign, you can help raise awareness and work with others to create positive change in your community. You don't have to organize a protest or stage a sit-in; you simply have to come up with a plan that's tailored to your community's and campaign's needs.

Here are some steps you can follow to help you get started:

1. **Choose an issue relevant to your community that's linked directly to climate change.**

   For example, perhaps your town doesn't have anti-idling bylaws, and you think it should. By taking on this issue, you can help fight climate change and protect kids who have asthma.

2. **Contact local organizations that are connected to or working on your issue, and ask them for resources and advice.**

   Toss a wide net — check out community service groups, youth groups, environmental groups, and so on.

3. **Find a meeting space.**

   Look for somewhere free or inexpensive, such as a public library or a community hall. Or host a videoconferencing meeting if you have a list of people already interested.

4. Select a date for your meeting.

Choose a time that you think works with most people's schedules. Also, you may want to offer child care. And think about accessibility for the physically challenged.

Give people advanced notice so that they can make arrangements to attend the meeting.

5. Reach out to the community to get as many people as possible to attend the meeting.

Send event information to radio stations and newspapers. Have friends put up signs and send out invitations. Send mass emails. Set up a group on Facebook. Don't be shy about getting the word out.

6. Speak your piece at the meeting, and then ask everyone else to talk.

Decide on your concrete goal, determine an action plan, and make sure that everyone has something to do when they leave.

Good luck, and let us know how it goes!

Spreading the Word

Believe it or not, letters to your elected representatives make a difference. Post-cards might catch someone's attention, and petitions do sometimes get noticed. But letters and emails are by far the most effective. And hard-copy letters in the mail still carry more weight than emails. Politicians are eager to know what the people think.

Don't worry about composing a long or deeply profound letter. Use the letter to express your concerns and ideas as simply as possible. To get more bang for your letter-writing buck, copy the letter and email it to other local, regional, and national politicians. You can even request to meet with a politician to discuss what the government is doing about climate change.

You can expand your outreach beyond political leaders. Writing to people within your industry, your business, or your children's school, or even writing to your own boss, can lead to profound change.

Whether you love to talk (which we do!) or hate to, you can spread the word. Formal presentations can effectively get the message on climate change across to your family, friends, co-workers, and community.

You can either give a presentation yourself or ask someone to come in and give it for you. Al Gore's campaign and NGO called The Climate Reality Project can show you how to present the same slide-show presentation that Gore gives in *An Inconvenient Truth.* So far, they've trained 20,000 activists in 152 countries. The Project's always planning more training sessions, so check their website (www.theclimateproject.org) for updates. From the site, you can request a presentation for your community or group.

You can also contact any local organization that's working on climate change to request a presentation. Who knows — maybe, after you get some pointers from the local organization, you can give the presentation yourself?

This kind of work by individuals on the ground can inspire and mobilize others to do the same thing. This adds a real multiplier effect to your work — when enough people are engaged, changes do happen. You can read more about this kind of social disruption in Chapter 17.

Chapter **19**

Ten (Plus Three) Inspiring Leaders in the Fight

Thousands of people around the world are working on climate change issues and making a difference. The ten we profile in this chapter are the cream of the crop. But, hey, remember — there's always room for one more!

We divide these leaders into these categories:

» **Politicians:** Around the world, politicians are struggling to decide how to address global warming. Despite facing budgetary challenges, resistance from industry, and a public concerned with countless other pressing issues, some leaders have managed to keep the climate crisis atop their agendas and are making positive changes.

» **Wordsmiths:** When it comes to fighting climate change, the pen is definitely mightier than the sword.

» **Activists:** People worldwide recognize global warming as a major issue thanks to the efforts of activists around the globe who are constantly reminding the media, the government, and the public about the climate crisis.

- >> **Scientists:** The scientists got the ball rolling with regard to climate change awareness, and their research continues to lay the groundwork for the world's understanding of climate change. Without the dedicated work of researchers, the world wouldn't know about the effects that our actions have on the planet.

- >> **Business leaders:** The private sector has a big role to play. You can find climate concerned leaders everywhere, but CEOs are often constrained by pressure to keep turning out value for shareholders (see Chapter 17 for more). Still, when business leaders speak, they reach people who aren't as impressed with scientists and activists.

Arnold Schwarzenegger

Past leadership has come from across the political spectrum. U.S. Republican Arnold Schwarzenegger went from tough guy movie star to climate fighter. California hit its 2020 target of reducing its GHG back to the level of its 1990 emissions. Thanks to creative flexible regulations and policies from the California Air Resources Board and carbon pricing through cap and trade with other jurisdictions, California is on track to reduce emissions to 40 percent below 1990 levels by 2030.

REMEMBER

Even though he's now a former politician, the former governor of California remains active in calling on the current crop of politicians to show courage. The Terminator takes to social media to keep making his point. His post-COP26 political leadership is keeper — look for it on YouTube.

Angela Merkel

Another retired leader from the conservative side of politics is former German Chancellor Angela Merkel. For decades, she played a key role in the G7 and in the European Union in pressing for meaningful climate action. Like Schwarzenegger in the previous section, Merkel can point to measurable change. Germany thumps California's record with a nearly 36 percent reduction against 1990 levels achieved early — by 2019. Germany's target at COP26 was to reduce to 55 percent below 1990 levels by 2030.

Mia Amor Mottley

The politicians of today showing climate leadership are those with the most to lose.

At COP26, Barbados Prime Minister Mia Amor Mottley emerged as a global leader. She made solid proposals and spoke with force and clarity. "Two degrees (3.6 degrees F) is a death sentence for us," she said.

"Our world stands at a fork in the road; one no less significant than when the United Nations was formed in 1945. But then the majority of countries here did not exist; we exist now. The difference is we want to exist a 100 years from now."

Mottley called for the International Monetary Fund (IMF) to use the mechanism used for COVID-19 to be applied to Climate Finance. Called Special Drawing Rights (SDRs), $650 U.S. billion was mobilized by IMF through SDRs for COVID. Her call for an annual drawdown of $500 billion in SDRs for carbon-cutting projects to hold to 2.7 degrees F 1.5 degrees C was well-received. Talks will continue to see if the United States and the IMF can make it happen.

Bill McKibben

Bill McKibben is a best-selling author of more than 15 books. He's also the founder of an international climate organization, 350.org. And one thing is clear: He has made a huge contribution to increased global public awareness of the climate crisis. His first book, *The End of Nature* (Random House), put forward the uncomfortable reality that the impact of humanity was so all encompassing that no part of the natural world existed outside humanity's influence. For many people, it was a first introduction to climate change.

George Monbiot

U.K.-based writer George Monbiot demands the attention of anyone interested in climate change. He' best known for his columns in *The Guardian* newspaper and his 2007 best-selling book, *Heat: How to Stop the Planet from Burning* (South End Press). His clear and compelling writings have brought attention to the perils of climate change. Monbiot has been praised for his forthright and demanding ideas for greenhouse gas (GHG) reductions, developing what appears to be a feasible plan for a 90-percent reduction in GHG emissions by the year 2030.

Out of the Wreckage: A New Politics for an Age of Crisis (Verso) is his most recent book. Ever the storyteller, he relates the human predicament as one needing a new story. In an affirmation of the essentially cooperative nature of humans, he sees humanity as "supreme cooperators." Far from the notion of "dog eat dog," cooperation is the secret of our success.

Sheila Watt-Cloutier

A long-time advocate for the rights of the Inuit people in the Arctic, Sheila Watt-Cloutier speaks out today on climate change. She is an Inuk woman and former head of the Inuit Circumpolar Council, representing Indigenous people throughout the circumpolar world, across national boundaries. The Indigenous peoples of the Arctic are already feeling its effects. Originally an advocate against contaminants in Arctic wildlife that the Inuit depend on, her focus expanded to climate change advocacy when she became the chair of the international Inuit Circumpolar Conference (the organization has since changed its name to the Inuit Circumpolar Council).

Watt-Cloutier has made the world understand that climate change is inextricably linked with the survival of Inuit culture and spirit. In 2007, Watt-Cloutier was nominated for the Nobel Peace Prize and garnered the Canadian Lifetime Achievement Award for this work. Her 2015 book, *The Right to be Cold* (University of Minnesota Press) made the case that global warming is a human rights violation.

Elizabeth Wathuti

A young woman from Kenya has been proving that one person can make a huge difference. Elizabeth Wathuti is the founder of Green Generation Initiative — a tree-planting organization for Kenyan youth. It was inspired by the late Wangari Matthai, who was the first environmentalist to ever win the Nobel Peace Prize. Like Wathuti, Matthai was a Kenyan tree-planter and climate campaigner. Matthai founded the Greenbelt Movement of Kenyan women planting trees. At COP26 in Glasgow, Wathuti took up the mantle of Matthai. She spoke to all the world leaders at a plenary that included U.S. President Joe Biden, UN Secretary General Antonio Guterres, UK Prime Minister Boris Johnson, and dozens of other powerful world leaders.

She said, "I have asked myself, over and over, what words might move you. And then I realized that making my four minutes count does not rest solely on me: My truth will only land if you have the grace to fully listen; my story will only move you if you can open up your hearts."

Greta Thunberg

Who would have ever imagined that a Swedish schoolgirl on the spectrum with autism could be one of the leading voices and launch millions of people to demand climate action? What started out as a small demonstration empowered a new generation of activists.

When *Global Warming For Dummies* came out in 2009, Greta Thunberg was 6. By 2018, at 15 years old, she started skipping school on Fridays — not a typical path to launch oneself into global activism. She sat outside her Stockholm school with her handmade sign *Skolstrejk för klimatet* (School strike for climate).

Media took note and before long, young people were joining Fridays for the Future. In September 2019, she led an enormous global climate action just days before a United Nations Climate Summit. In the world's largest day of climate action ever, on September 20, 2019, more than four million persons took part in more than 150 countries.

Her thinking was clear: If the actions of grown-ups was to deprive children of their future, why were they in classrooms? She made it clear to adults: "Our house is on fire. We have to act like our house is on fire." And her recent summary of all the millions of words and thousands of promises made by the powerful was clear and powerful: "Blah, blah, blah," she said. Without action, just blah.

Michael Mann

Michael Mann, PhD, has been a leading climate scientist for a long time. His work set out the rapidly rising warming gases in the atmosphere over a thousand-thousand-year period for what became known as the "hockey-stick" graph. He's a professor at Penn State University and a lead author in the "Observed Climate Variability and Change" chapter of the Intergovernmental Panel on Climate Change (IPCC) Third Scientific Report.

He has been the target of more personal attacks than most scientists solely because he makes a strong case that humanity must move away from fossil fuels rapidly. He has stood his ground: He's gone to court, sued for defamation, received apologies, and moved on. His book, *The New Climate War: The Fight to Take Back Our Planet* (PublicAffairs Books) details the 30-year campaign of disinformation from the fossil fuel lobby.

Scientists shouldn't have to put on a superhero cape to do their work. Unfortunately, a lot of them have had to do just that.

Katharine Hayhoe

Katharine Hayhoe, PhD, is a very rare scientist in that she is equally committed to doing her cutting-edge research in atmospheric science and to explaining it to the average person. Even a bit more unusual, she is an Evangelical Christian as is her husband who also is a pastor. Originally from Canada, she lives and works in Texas and does a new kind of missionary work — explaining why Christians need to take the climate emergency seriously.

Her 2021 book *Saving Us, A Climate Scientist's Case for Hope and Healing in a Divided World* (Simon and Schuster) combines all her areas of her expertise. Furthermore, her TED Talk has close to four million views — not your average scientist.

James Hansen

People around the world who value courage in telling the truth despite extraordinary pressure and threats applauded James Hansen, scientist and advocate for urgent solutions to climate change, for speaking up when the U.S. federal government tried to censor his work with NASA on climate change. This censorship is the subject of Mark Bowen's book *Censoring Science: Inside the Political Attack on Dr. James Hansen and the Truth of Global Warming* (Dutton).

Although Hansen works as a physicist on major climate change modeling, his work outside the office has had the greatest impact on the fight against global warming. He has made it a personal mission to communicate climate change to the public in the most clear-cut way possible.

Bill Gates

Bill Gates is known world-wide as a massively successful entrepreneur, co-founder of Microsoft, billionaire, and philanthropist who, with his former wife, founded the Bill and Melinda Gates Foundation. In his book, *How to Avoid a Climate Disaster: The Solutions We Have and the Breakthroughs We Need* (Alfred A. Knopf), he confessed that for years, he set aside the climate crisis as someone else's problem. Gates found his reason to be concerned about climate change through his work in the developing world, realizing how desperately much of the world needs access to electricity, and then realizing that electricity couldn't be fueled by carbon.

Gates sets it out in clear terms. Humanity emits 51 billion tons of carbon dioxide equivalents/year and globally human activities have to get to zero. And he sets out, equally clearly, how humans need to get there from here:

>> **Electrify every process possible.** Doing so is going to take a lot of innovation.

>> **Get that electricity from a power grid that has been decarbonized.** Doing so will also take a lot of innovation.

>> **Use carbon capture to absorb the remaining emissions.** More innovation is needed.

>> **Use materials more efficiently.** You guessed it, more innovation.

Nandita Bakhshi

Nandita Bakhshi is one of the top female executives in U.S. finance. She has been the chief executive at San Francisco-based Bank of the West for five years. *American Banker,* an industry publication, consistently ranks her among the most powerful women in U.S. banking, based on financial performance and her leadership on diversity and environmental issues.

Bakhshi's corporate biography points out that as CEO, she introduced policies that made Bank of the West "one of the few U.S. banks to restrict financing of activities deemed harmful to the planet." As CEO, she raised hackles in the energy industry by cutting off loans to fracking and oil sands companies.

Chapter **20**

Top Ten Myths about Climate Change Debunked

G lobal warming has been a topic of debate for decades, but the discussion has only recently permeated all countries and cultures. In some cases, the fossil-fuel industry has financed major public relations campaigns to promote doubt about the level of risk and degree of scientific consensus concerning global warming. Because of its efforts, a great deal of misunderstanding was created. In fact, if you ask some folks, they'll tell you that the whole "climate change thing" is nonsense. If you ask others, they know the crisis is real, but they're wracked with guilt, feeling it's their fault.

In this chapter, we tackle the ten most common myths in those two categories — and offer arguments that refute them.

Knowing That a Debate Doesn't Exist among Scientists

The idea that a big scientific disagreement exists regarding global warming is one of the most persistent and erroneous claims used to delay action. For more than a decade, the science has been settled. The Intergovernmental Panel on Climate Change (IPCC) has consistently reported broad agreement which, in the 2021 Sixth Assessment Report, rose to nearly 100 percent certainty. The IPCC underestimates the severity of the risk and overestimates how much time we have to avoid the worst of the coming climate crisis.

Here are the firm and well-confirmed results of the largest peer-reviewed scientific effort in the history of humankind:

>> The planet is and has been warming, and it will continue to warm for the foreseeable future.

>> The warming isn't happening because of natural factors alone; it's due to human activity — burning oil, coal, and gas, and destroying forests.

>> The impacts of the climate changes that are happening because of rising temperature are increasingly catastrophic causing billions of dollars in economic losses and loss of life.

>> Urgent global action is required to hang on to a planet that's hospitable to human society and prosperity.

Over time, fewer and fewer scientists are claiming these facts are in dispute, although the fossil fuel lobby wants to amplify voices of the dwindling few. And the news media is used to the idea that objective reporting includes presenting both sides, even when one "side" is a very small minority of opinion. A generation ago, cigarette companies used similar tactics, finding doctors to dispute that smoking caused lung cancer. Al Gore even uses an old cigarette ad that shows doctors recommending a particular brand of cigarette in the 2006 documentary *An Inconvenient Truth*. Refer to Chapter 16 for more details how the media has covered climate change.

Recognizing That Human Activity Has Caused Current Global Warming

You may hear global warming skeptics say that weather goes in cycles and the recent warming trend is just that: a trend that will correct itself. It's true that natural climate variation exists and always has. Scientists have records of wide temperature variations over time, ranging from the temperature lows of the Ice Ages to many millions of years ago when the Arctic was a swampland. A combination of temperature, greenhouse gas (GHG) emissions, plate tectonics, and the sun affect the climate. Of course, millions of years ago, Planet Earth wasn't hospitable to human life. Humans showed up more recently (as Chapter 1 explains). Humans have had the Goldilocks planet — not too hot, not too cold — just right!

But today's quickly rising temperatures are unlike any previous recorded changes.

REMEMBER

The threats from global warming doesn't come simply from things being warmer. The threat comes from destabilization of really large systems — like ocean currents, ocean acidification, and Arctic ice. Whole climate systems are being thrown out of whack — from the Gulf Stream to the jet stream and to polar vortexes and atmospheric rivers. Large system changes result in positive feedback loops, leading to tipping points where whole systems suddenly shift into entirely new behaviors.

There is a level of warming that at some future point could result in unstoppable, self-accelerating warming. Humanity must avoid if we're to have a functioning society as a human civilization.

With such big threats, it's only human nature to cling to any plausible explanation to avoid reality. Some people use what seems like a commonsense approach to explain away global warming. They point out that humans didn't keep good temperature records thousands or even hundreds of years ago. True. But the data extends further back than the approximately 140 years that people have been keeping temperature records. Scientists can make reliable estimates about the baseline temperature trends by using indirect measurements from sources such as ice cores from glaciers and tree rings from ancient forests.

Science that bases its conclusions on concrete evidence — from ice core samples to melting glaciers — makes a clear case that the warming of the modern world is well beyond natural variation. The atmospheric concentrations of the GHG carbon dioxide are now about 47 percent above the highest levels in the last 800,000 years (based on direct measurements from ice cores), and have likely never been this high for more than 20 million years. Or put another way, no human being in the history of the species has ever breathed air with as high concentrations of carbon dioxide as they breathe now.

Looking into the Danger of Increased Carbon Dioxide Concentrations

Carbon and temperature aren't always linked in their historical records. But it doesn't make sense to take a bit of the truth, such as this fact, and use it to create the impression that no link has been proved between human-caused emissions of carbon dioxide and changes to global temperature (and hence the climate). Sure, over a geologically long period of time, carbon dioxide levels and temperature aren't always in lockstep. Carbon dioxide is only one of many GHGs and only one part of the climate equation. However, carbon dioxide is the GHG that's playing a lead role now.

Carbon dioxide is a powerful warming gas. About 280 parts per million (ppm) of carbon dioxide kept the planet livable for millennia. That many parts per million is like a drop in a swimming pool. When the concentration of carbon dioxide increases, the atmosphere heats up significantly. That said, an increase in carbon dioxide isn't the *only* thing that's ever heated up the atmosphere.

REMEMBER

Carbon dioxide *is* an essential element of life, but it's just one of many factors that influence the world's temperature. Overall, however, the more carbon dioxide you emit into the atmosphere, the warmer the Earth gets. Chapter 3 delves deeper into what you need to know about carbon dioxide.

Discovering the Truth about Sunspots

A number of factors influence the temperature on the planet, including *solar flaring* (or sunspots) and *radiative forcing* (the impact of the sun's activity, which varies, on the climate). The IPCC scientists reviewed the literature about the possibility of sunspots and changes in the sun's activity affecting the planet's climate. They estimate that the warming effect from increases of GHG in the atmosphere is more than 50 times greater than the relatively minor effect of solar irradiance changes.

Understanding That Scientists Don't Exaggerate to Get More Funding

Doubters suspect that scientists overstate the urgency of global warming in order to obtain more grants. The notion that the world's leading scientists and scientific academies fudge the evidence and put themselves at odds with the most powerful corporations on Earth for money is pretty far-fetched. If anything, the environmental community and scientists have been too cautious in expressing concerns.

The anticipated changes documented in the early IPCC reports, starting in 1990, were all based on what would happen if the carbon dioxide emissions doubled, which led some to think that the only risk was if the concentration of GHGs doubled to 550 ppm. Now, scientists increasingly understand that the risk to societies and ecosystems is unacceptably high at levels far below 550 ppm.

REMEMBER

Increasingly, scientists are calling for stabilization of carbon dioxide emission below 425 to 450 ppm — the atmosphere is at 412 ppm already, and rising at approximately 2 ppm per year. The increasing concentration of GHGs has resulted in average global temperature increase of 2 degrees F (1.1 degrees C). That doesn't sound like a big difference — but remember — it's a temperature expressed as a global average. That little number is the reason that extreme weather events are getting more frequent and more dangerous.

The IPCC says that globally all countries need to cut carbon dioxide emissions by 45 percent by 2030, leading to net zero carbon by 2050.

Grasping the Misconceptions about the Science of Global Warming

You may hear people talking about new technologies that can save the world — clean coal, a hydrogen economy, or carbon capture and storage. Stick around long enough, and you even hear really wild sci-fi solutions, such as adding iron oxide to the ocean or putting mirrors in outer space. Studies show that many of these technologies, especially forms of *planetary engineering* or *geo-engineering* — in other words, large-scale re-engineering of the planet — would make things worse.

Still, high-tech solutions are available, and you're bound to see new breakthroughs. Each potential solution, especially the proposed use of hydrogen and fuel cells, may play an important role in the future, but the Earth doesn't have the luxury of time. We can no longer postpone needed action.

Solutions that sound too good to be true are probably just that — too good to be true. No solution allows humanity to have its cake and eat it, too. All around the world, emissions must go down — fast. Really, everyone needed to reduce those emissions yesterday.

Thinking It's All Your Fault

This myth is sneaky. As co-author of the first edition, Elizabeth bought into it too. People could all play their part and reduce GHG emissions. All of us in our everyday decisions — taking a bus or a car or putting on a sweater or turning up the heat — were contributing to global warming. And those big polluters? They're only responding to consumer demand.

Governments loved this myth. Politicians, particularly in the United States and Canada, said things like, "Is the public really ready to take public transit and leave the car in the driveway?" Politicians used public readiness as an excuse for inaction. That never happened when using regulatory tools to protect and heal the ozone layer or to protect public health by banning lead in gas.

REMEMBER

Then 2020 and the year of COVID hit. Around the world, airplanes were grounded and cars stayed in the drive. What was the overall result? GHG emissions did drop all around the world, but only by 6.4 percent compared to 2019. The truth is systems currently hard-wired to fossil fuels primarily drive global warming. Electricity grids powered by coal. Trucking and shipping all rely on fossil fuels.

Without massive decision-making at the level of national governments and big corporations, the Earth won't avoid the worst.

Does that mean there is no point to individual action?

Not at all. Each individual action taken reminds everyone they aren't powerless! Any time you take some action, and a friend, family member or neighbor hears or reads about it, you are a part of growing the democratic strength that will move governments and industry faster than they would otherwise like to. That's vital work.

TIP

You should never miss a chance to talk to friends and neighbors, colleagues and bosses, anyone really, about how you see the climate emergency and what action you can take, and perhaps more importantly, what action you want industry and government to take.

Considering You Can't Do Anything about It

Defeatism is a big threat to the Earth's survival. It runs a close second to apathy. Making citizens feel powerless is one way that polluters get away with damaging everyone.

Knowing everyone can all make a difference — in big and little ways — is key. So, if you remember to turn out the lights in an empty room, that makes a difference. If everyone remembered, it would make a huge difference. But that alone won't arrest the climate emergency at a livable level of disruption.

TIP

As you (and we) shrink your carbon footprint, in order to make a difference, we all have to expand our political footprint. Every little step taken increases the sense of personal engagement — and personal power.

Everyone does have individual power and agency — particularly those individuals living in wealthy, industrialized democracies. Everyone has the power to improve their own carbon reduction efforts, educating other people in the community, while pressing all levels of government to slash emissions, boost tree-planting and other activities that restore nature, and put in place meaningful protections for workers. Just as earlier generations did in war time, with an "all hands on deck" attitude, there is nothing that can't be done.

Pinning the Blame on Developing Countries Isn't Realistic

Another dangerous myth is that industrialized countries (like the United States and Canada) acting to reduce emissions won't make any difference because the developing nations are totally unregulated.

Under the Kyoto Protocol, countries considered *developing* (including China, India, and Brazil) didn't take on firm targets to reduce emissions. That's no longer the case. Under the Paris Agreement, all nations have taken on commitments, described under the terms of Paris as NDCs — *nationally determined contributions*.

As of now, China has made more investments in renewable energy than any other country on earth. Although China is still a big polluter, so are the United States and Canada. Every country needs to do more — and do it as fast as possible — together. Refer to Chapter 12 for more information.

Living with Climate Change — and Doing Something about It

That humanity can adapt isn't really a myth. Within limits, people *can* adapt. But humanity's adaptability doesn't mean people can ignore the demands to reduce greenhouse gases. Some areas of the world are already so hot that humans can't live there — you don't find cozy habitations in Death Valley or the Lut Desert in Iran or vast areas of Queensland in Australia.

People can adapt to the impacts of climate change at the lower end of impacts. The Paris Agreement doesn't speak in terms of parts per million, but in terms of limits to global average temperature increase. Small increases in global *average* temperature don't mean that temperature increases everywhere are small. Some of the hottest places on Earth are heating up more than the average. That means that more places simply won't be survivable for humans. The people that live there now will have to go somewhere else. That' isn't adaption; it's climate flight.

All countries on Earth committed under the 2015 Paris Agreement to hold global average temperature increase to no more than 2 degrees C (3.7 degrees F). That translates to roughly no more than 420 ppm.

But humanity doesn't know where the new concentrations of GHG will level off. What will be the new normal? If the world can reduce GHG quickly enough, humanity could hold that new normal to 400 to 425 ppm. Many scientists are arguing that carbon must be brought back down to no more than 350 ppm, which would require as yet undeveloped techniques to actually withdraw carbon from the atmosphere (more on this in Chapter 13).

Humans living on low-lying islands and along low shorelines can't adapt to rising ocean levels above the ability to build dikes and pump water out of the lowest areas. Adaptation to having your home under water means moving to somewhere else.

But the Earth isn't yet locked into levels of climate disaster beyond the ability to adapt. Humanity does needs to adapt to the damage we've already done. But the focus has to be on reducing emissions to keep climate change within the limits beyond which adaptation just isn't possible.

Chapter **21**

Ten (Plus One) Online Climate Change Resources

Global warming has been on the scientific and political agenda for nearly 40 years, so it's no surprise you can find a lot of information about it online. But sorting through all the information and verifying its veracity is challenging. You need to have some critical external verification that the information is reliable.

Nowadays, the Internet is chockfull of solid research from good accredited sources. Although some disinformation is still floating around, most of the climate change deniers have stopped their ill-informed ranting. Be careful around social media sites, though — there you can still find people who think that the earth is flat, that aliens live among us, or that human-caused climate change doesn't exist.

This chapter points you to some of the best sites, whether you want to explain to your kids about climate change, need a place to start for your own research, or want to know as much as you can about the issue.

National Geographic

A great site with information for parents and kids comes from National Geographic at www.natgeokids.com/. Just type in "climate change" in the search window. This site also includes the following content:

>> **Games:** Each game shows kids how to make the best choices about climate change and the environment.

>> **Homework help:** This site offers interactive information to help kids with homework related to climate change, energy, land use, and water, among other topics.

>> **Content for teachers:** With resources for in-class curriculum, book lists, and resources, teachers find the site a dream.

Three Other Kid-Friendly Sites

Depending on their interests, active young person looking for a way to make important impacts can find it online. Here are some other kid-friendly sites that we highly recommend:

>> **Fridays for the Future:** Increasingly, climate organizations are developed and run by young people to increase their own advocacy skills. Fridays for the Future is an international youth-led movement with effective tools for activists available at https://fridaysforfuture.org/.

>> **NASA:** You can help the kids in your life understand the environment and climate change by sourcing climate education from reliable sources. NASA has a great website for climate education geared for kids. Check out https://climate.nasa.gov.

>> **Youth Climate Lab:** Based in Canada, and operating globally, Youth Climate Lab (www.youthclimatelab.org/) is another effective youth-led climate organization. Youth Climate Lab is more structured around skills development and youth-led projects.

Intergovernmental Panel on Climate Change

Created in 1988 by the World Meteorological Organization and the United Nations Environment Program (UNEP), the Intergovernmental Panel on Climate Change (IPCC) (www.ipcc.ch/) is a team of scientific experts who, among other duties, writes reports for the United Nations. These experts, appointed by their governments, look at all the peer-reviewed science in the world relating to global climate change and synthesize it into assessment reports. On their site, you can access their numerous reports, which cover the science behind climate change, its effects, and potential solutions to it.

TIP

If you want the detailed scientific background to climate change, check out the IPCC's full reports; if you're just looking for the big picture, we recommend the "Summaries For Policymakers."

Climate Analysis Indicators Tool

This site (http:/cait.wri.org), offered by the World Resources Institute, enables you to create graphs that compare greenhouse gas (GHG) emissions by country, by source of emissions, by carbon dioxide per person by country — or by pretty much any other combination you can think of, actually. You need to sign up in order to use the tool, but it's free.

You may find the site a little intimidating at first glance because it's not geared to the general public. Just dive in and use the tool, however, to become more comfortable.

Canada's Environment and Climate Change

Canada's simple climate change website, www.canada.ca/en/environment-climate-change.html, offers a brief overview of climate change, suggests actions for you to take at home, and provides access to speeches about climate change made by government officials. By digging around the site a bit, you can also find the following information:

>> Reports on Canada's GHG emissions and commitments

>> How to apply for incentives and rebates that apply to activities such as renovating your home, upgrading your car, or doing pretty much anything else that improves your personal energy efficiency

>> Opportunities for citizens to provide advice on climate policies

>> Examinations of climate change and its relation to agriculture, energy, and transportation in Canada

The U.K.'s Climate Challenge

The British government's climate change site (www.climatechallenge.gov.uk) is our personal favorite because it's clear, direct, and to the point. At this comprehensive website, you can

>> Bone up on the basics of global warming

>> Calculate your own carbon dioxide emissions

>> Look into the climate change projects, initiatives, and policies of the U.K. government

>> Get involved in local projects

Environmental Protection Agency

The U.S. Environmental Protection Agency's site offers a bunch of useful information. This excellent website was shut down for a while, but is now back up and running. By browsing around www.epa.gov/climatechange, you can find the following:

>> The government's climate policy

>> The effects of climate change on health and the environment

>> Links to your state and local governments, as well as their action plans

The International Energy Agency

If you get energized thinking about the possibilities for power that exist beyond fossil fuel, check out the website for the International Energy Agency (IEA), www. iea.org. The IEA acts as an energy consultant for its 27 member countries, which include the United States, Canada, Australia, and the United Kingdom. Originally created during the energy crisis of the mid–1970s to deal with oil supply emergencies, the IEA's focus has changed with the times, and it now deals with all sources of energy.

REMEMBER

The IEA's site is for the ambitious. If you want to know the ins and outs of the global energy world, this is the place to find it. Throughout the entire site, you can find information on the following:

>> **All forms of energy:** The IEA offers a wealth of reports about energy sources such as renewable sources, natural gas, and oil. Select the energy of your choice from a list of links under the Topics tab, and the site displays information on publications, programs, workshops, and even contact information if you have questions.

>> **Specific energy issues:** The IEA provides information about how energy is relevant to many different subject areas, such as sustainable development, and emissions trading.

>> **Country-by-country data:** The IEA site indicates how much energy each country uses, where this energy goes, and the source it comes from. You can select a country name from a list or click a region on the site's map. Each country's page contains links to the amount of oil and type of renewable energy used by that country, as well as the country's energy policies.

>> **How the IEA works:** The site explains how the IEA works with energy producers, industrialized countries with high energy demands, and developing countries (such as India and China) to try to make energy supplies available and sustainable in the face of climate change.

The IEA has become a leading voice in the pricing of carbon and for the end of fossil fuel subsidies. It has maintained accurate information on the rapid shift to renewables, and it also operates a separate site about carbon capture and storage, a topic that we cover in Chapter 13.

Gateway to the UN System's Work on Climate Change

The United Nations is a giant organization, with an array of suborganizations, many of which are working on global warming issues. A large number of UN departments have climate change projects. These UN sites demonstrate the connections between climate and other important international issues, such as food, human settlements, and economics. Go to www.un.org/climatechange to explore the goings-on of these organizations.

Beyond the links to these programs and organizations, this site offers the following:

>> **International climate change projects:** A list of incredible climate change projects, categorized by country and led by various UN departments.

>> **Resources for children and youth:** Games, programs, and event listings relevant to young people.

>> **General climate change information:** Fact sheets that contain information regarding climate science, politics, and actions to take.

>> **The U.N Framework Convention on Climate Change (UNFCCC):** This is the context within which the Kyoto Protocol and Paris Agreement were negotiated. Check it out for constantly updated information on the work of the UNFCCC.

Index

Bank of England, 262

banking industry, 261–263, 306

Barclays, 259

batteries, 308

Belarus, as an economy in transition, 192

Below 2 Degrees Coalition, 275

Bicycle Kingdom, 207

bicycles, Municipal Initiatives for, 171–172

Biden, Joe (US president), 198, 332

Bill and Melinda Gates Foundation, 335

biodiesel, 244

Biodiversity Initiative, 167

bioethanol, 244

biofuels
  about, 242
  defined, 241
  supplying in farming and forestry industries, 265

biomass, 242–243

blogs, for climate change, 293–294

bogs, ecosystems of, 129–130

Bolsonaro, Jair (Brazil president), 208

Bonn, Germany, 263

books, on climate change, 294–296

boreal forests, 131–133

Borrelia burgdorferi, 140

Bowen, Mark (author)
  Censoring Science: Inside the Political Attack on Dr. James Hansen and the Truth of Global Warming, 334

Boykoff, Jules (academic), 286

Boykoff, Maxwell (academic), 286

Bradley, Brendan (author)
  ESG Investing For Dummies, 324

Brazil
  climate initiatives in, 183–184
  Curitiba, 208
  as a developing country, 203
  hydropower in, 239
  as a Non-Annex 1 country in U.N. Framework Convention on Climate Change, 192
  progress in, 203–204
  sugarcane in, 242

British Petroleum (BP), 255, 260

Browne, John (CEO), 260

Bruggemann, Michael (researcher), 286

Building Research Establishment and Environmental Assessment Method (BREEAM), 258

buildings
  certifying new, 257–258
  climate change and, 146
  efficiency in, 225
  energy alternatives for, 257
  reducing heating and cooling, 256
  regulations for, 178–179
  restrictions on coastlines, 169
  using energy in, 79–80

Burtynsky, Edward (artist), 291

Bush, George H.W. (US president), 189

business and industrial solutions
  about, 249–250
  adaptation for, 255
  assets and, 262–263
  banks and, 261–262
  carbon market, 254–255
  certifying new buildings, 257–258
  conserving energy, 251–252

corporate nonsuccess stories, 259–262

corporate success stories, 258–259

energy alternatives, 257

farming, 264–269

forestry, 264–269

greener buildings, 256–258

insurance industry, 263–264

law firms, 264

liabilities and, 262–263

manufacturing, 250–254

processing, 250–254

reducing heating and cooling, 256

support from professional service sector, 262–264

using energy efficiently, 252–253

Business Environmental Leadership Council (BELC), 276

business leaders, as leaders in climate change, 330

C

C-40 Cities Climate Leadership Group, 170

California
  carbon taxes in, 181
  climate initiatives in, 183–184
  emissions trading in, 160–161

California Air Resources Board (CARB), 184

California Global Warming Solutions Act (2006), 184

Canada
  adaptation funding and, 218
  as an Annex 1 country in U.N. Framework Convention on Climate Change, 192

CAN, 277

Clean Development Mechanism (CDM), 163, 211, 218–219, 253, 267

clean energy, in China, 206

climate, controlling in your home, 96–98

Climate Accountability Act, 158, 177

Climate Action Network (CAN), 277

Climate Analysis Indicators Tool, 347

Climate Cent Fund, 185

Climate Challenge, 348

climate change. See also specific topics

about, 7–8

history of, 9–12

living with, 344

myths about, 337–344

online resources for, 345–350

tips for fighting, 317–328

climate computer models, 55

Climate for Change, 154

Climate Group, 275

climate justice, 280

climate pacts, ratifying, 197–198

Climate Savers, 276

The Climate Reality Project, 328

clinker, 82

Clinton, Bill (US president), 235

Clooney, George (celebrity), 292

closed loop system, 253–254

cloud cover, as a cause of global warming, 9, 49–50

coal plants, 67–68, 243

coastlines, building restrictions on, 169

co-firing, 243

co-generation, 226–227

combustion system, 93

compact fluorescent light bulbs (CFLs), 227

Conference of the parties (COP), 193

conserving energy, 251–252

consumer goods, incentives for, 160

contaminated drinking water, worsened by global warming, 141

Convention on Biological Diversity, 189

cooling

in homes, 97–98

reducing in buildings, 256

cooperative approaches, to emissions trading, 163

COP3, 163, 194

COP15, 194, 217

COP19, 219

COP21, 163, 194

COP22, 163

COP25, 163

COP26, 68, 163, 195, 206, 208, 211, 213, 312, 330, 331, 332

Copenhagen Accord, 195

coral, in oceans, 126–128

corn-based ethanol, 241

corporate cooperation, NGOs and, 275–276

costs

of disruption, 304–306

of global warming, 145–149

countries. See also specific countries

climate initiatives in, 185–186

in U.N. Framework Convention on Climate Change, 192–193

what they can do, 214–219

COVID-19 pandemic, greenhouse gases (GHGs) and, 92, 94

cows, 269, 322

crabeater seals, 136

Crichton, Michael (author)

State of Fear, 296

crop diversification, 168

crude steel, 82

currents, in oceans, 240

D

daily living, energy-efficiency and, 319–322

dams, 240

David Suzuki Foundation, 282

The Day After Tomorrow (film), 288–289

de Pencier, Nicholas (filmmaker), 291

decarbonization, 224

deforestation

about, 13, 87–88

in Brazil, 210–211

degrees, of temperature, 55

Dengue fever, 139–140

Denmark

climate initiatives in, 185

governments in, 177

Paris Agreement and, 194–195

wind energy in, 233–234

Department of Defense, 149

Desmog Blog (blog), 294

developed countries

about, 147–149

challenges faced by, 202–204

what they can do, 216–219

developing countries

about, 23, 147–149, 343

challenges faced by, 202–204

what they can do, 214–216

developments

about, 201

challenges faced by developed/developing nations, 202–204

endangered nations, 217

Harrelson, Woody (director), 291

Hayhoe, Katherine (author), 313

Saving Us, A Climate Scientist's Case for Hope and Healing in a Divided World, 334

HDR, Inc., 259

heat

about, 116–117

combing with power, 226–227

heat domes, global warming and, 14–15

Heat: How to Stop the Planet from Burning (Monbiot), 295, 331–332

heating

in homes, 96–97

reducing in buildings, 256

heatwaves, 17, 148

"high confidence" (IPCC), 112

highways, 146–147

"hockey-stick" graph, 333

Hollywood, climate change and, 288–293

homes

as biomass sources, 243

energy-efficient, 319

hot spots, 238

household energy, 95–98

How to Avoid a Climate Disaster: The Solutions We Have and the Breakthroughs We Need (Gates), 335

Howard, John (politician), 158

Hungary, as an economy in transition, 192

Hurricane Dorian, 114

Hurricane Florence, 114

Hurricane Harvey, 114

Hurricane Juan, 114

Hurricane Katrina, 148

hurricanes, 113–114, 148

hydraulic fracturing (fracking), 41

hydro energy, 24

hydrocarbons, 67, 244–245

hydrofluorocarbons (HFCs), 42

hydropower, 233, 239–240

I

I Am Greta (documentary), 291

Ice Age 2: The Meltdown (film), 291

ice melt, global warming and, 21

Iceland, geothermal energy in, 238

ICLEI - Local Governments for Sustainability, 170

icons, explained, 3

Idaho, methane in, 245

illnesses, global warming and, 19

impoundment systems, 239

incandescent light bulbs, 227

incentives, government, 160

India

as a developing country, 202, 203

as a Non-Annex 1 country in U.N. Framework Convention on Climate Change, 192

progress in, 203–204

Indigenous people, NGOs and, 277–278

indirect energy, 95

Indonesia, geothermal energy in, 238

industrial solutions. *See* business and industrial solutions

industrialized countries. *See* developed countries

industries

adaptation for, 255

capturing carbon dioxide during emissions, 228–229

dependence on, 25

disruption and industrial energy, 306

efficiency in, 225

relationship with NGOs, 275–277

using energy in, 81–84

influencers, NGOs and, 281

information sources, as sources for news stories, 287

infrastructure, 146–147

Institute of Science in Society, 126

insulation, in buildings, 256

insurance industry, 263–264

Intel, 259

Interface, 259

Intergovernmental Panel on Climate Change (IPCC), 16, 48, 72, 87, 93, 103, 111, 112, 123, 139–140, 174, 198–200, 228, 251, 265, 286, 338, 347

intermodal transport, 172

internal combustion engine (ICE), 224, 320

International Energy Agency (IEA), 69, 78, 85, 173, 207, 224, 225, 231, 300, 302, 349

International Monetary Fund (IMF), 175, 331

International Renewable Energy Association (IRENA), 174–175, 232, 300, 301

International Union for the Conservation of Nature (IUCN), 153

Internationally Transferred Mitigation Outcomes (ITMOs), 163

Inuit, 151

investing, 324

iron, manufacturing, 82

iron ore, 82

irradiance cycles, 49

islands, small, effects of global warming in, 20

It's Getting Hot In Here (blog), 294

metals, manufacturing, 82–83

methane (CH$_4$)

about, 40–41

in garbage, 244–245

Miami-Dade County, FL, climate goals of, 166

Milankovitch cycles, 49

mining companies, 255

mitigation

defined, 192

by developing countries, 214

modalities, for emissions trading, 164

modeling, carbon and, 53–54

Monbiot, George (author)

Heat: How to Stop the Planet from Burning, 295, 331–332

Out of the Wreckage: A New Politics for an Age of Crisis, 332

Montreal Protocol, 44–45, 197

Moore, Michael (filmmaker), 291

Mottley, Mia Amor (Prime Minister), 331

mountain glaciers, 106–107

movements, of oceans, 126

movies, climate change in, 288–291

N

Nappalak, Naalak (elder), 151

NASA, 346

National Environmental Friendly Vehicles project, 207

National Geographic, 346

National Petroleum Council (NPC), 69

Nationally Determined Contributions (NDCs), 179–180, 193, 196

natural disasters

about, 103

effect of natural disasters caused by climate change on women, 153–154

flooding, 111–113

forest fires/wildfires, 114–116

freshwater contamination, 113

heat, 116–117

hurricanes, 113–114

melting glaciers, 106–108

oceans, 109–111

positive feedback loops, 117–119

rainfall, 111

rising sea levels, 104–106

storms, 113–114

natural gas, 71–73

Nature Climate Change (magazine), 110

nature-based climate solutions (NBCS), 124

Nelson, Willie (artist), 292

net zero, 298, 300

New Zealand

as a developed country, 202

effects of global warming in, 19–20

news, climate change and, 285–288

The New Climate War: The Fight to Take Back Our Planet (Mann), 333–334

Nickels, Greg (mayor), 182

Nike, 274

nitrogen, as a component of air, 30

nitrous oxide (N$_2$O), 42, 43

Non-Annex 1 countries, in U.N. Framework Convention on Climate Change, 192

nongovernmental organizations (NGOs)

about, 251, 271–272

climate justice, 280

examples of, 282–283

generation and, 278–279

getting involved, 280–283

jobs with, 325

relationship with industry and government, 275–277

role of, 272–278

nonmarket cooperation, to emissions trading, 164

nonrenewable energy, 245–247

North American Commission for Environmental Cooperation, 85

North Dakota

emissions from, 230

fossil fuel infrastructure in, 277

Northern Europe, carbon taxes in, 182

Northern Hemisphere communities, effect of global warming on, 150–152

Northwest Passage, 108

Norway

carbon taxes in, 182

emissions from, 230

governments in, 177

nuclear fission, 245

nuclear power, 245–247

O

Obama (US president), 158

O'Brien, Karen (author)

about, 313

You Matter More Than You Think, 296

oceans

about, 33–34, 109–111

ecosystems of, 125–129

power of, 240–241

oil production

about, 68–71, 83–84

capturing carbon dioxide during, 228–229

"1.5 to stay alive," 298, 300–303

online resources, for climate change, 345–350

onshore wind energy, 233

Oregon, carbon taxes in, 181

organic materials, 244

Organization of Petroleum Exporting Countries (OPEC), 74

organizers, NGOs and, 281

Orlowski, Jeff (filmmaker), 290

Out of the Wreckage: A New Politics for an Age of Crisis (Monbiot), 332

outbound tourism, 94

outbreaks and diseases

  about, 138–139

  Dengue fever, 139–140

  Lyme disease, 140

  worsened by global warming, 140–142

oversimplification, of news sources, 287–288

Oxfam, 280

oxygen, as a component of air, 30

ozone depleters, 44–45

ozone layer, 19

P

paper and pulp industry, 84, 243, 253–254

Paris Agreement

  about, 23, 158, 170, 179–180, 182, 193, 253, 298–299, 343, 344

  adding flexibility, 196–197

  ratifying climate pacts, 197–198

  setting targets, 193–196

partners, developed countries as, 218–219

passive solar energy, 236

peak oil, 70

Peduto, Bill (mayor), 170

peer-reviewed science, 199

Pembina Institute, 276

people

  about, 137–138

  agriculture, 143–144

cost of global warming, 145–149

effect of global warming on, 137–154

outbreaks and diseases, 138–142

types, 149–154

perfluorocarbons (PFCs), 42, 83

permafrost, 20, 41, 56

personal vehicles, initiatives for, 173–174

pests, forest, 132–133

Philippines, geothermal energy in, 238

photosynthesis, 32, 34–36

photovoltaic energy, 235, 308

phytoplankton, 126, 127

Pittsburgh G-20 Summit, 175

Planet of the Humans (documentary), 291

planetary engineering, 341

plants

  about, 34–36, 121–122

  ecosystems, 122–124

  energy from, 241–242

  forests, 130–133

  underwater, 124–130

Poland, as an economy in transition, 192

polar bears/animals, 135–136

polar regions, effects of global warming in, 20–21

politics

  changes in, 309–310

  governments and, 158

  politicians as leaders in climate change, 329

pollution, lowering from coal plants, 67–68

poop, seeds in, 124

population growth, greenhouse gases (GHGs) and, 73–74

positive feedback loops, 117–120

potable water, 146

poverty, people in, effect of global warming on, 152

power

  combing with heat, 226–227

  disruption and, 305

precautionary principle, 190

prehistoric evidence, for carbon, 52–53

private research, 165

procedures, for emissions trading, 164

production

  of electricity, 78–79

  in oceans, 126

professional service sector, support from, 261–264

provinces, climate initiatives in, 183–184

public research, 165

public speakers, NGOs and, 282

public transportation, initiatives for, 172

pulp and paper industry, 84, 243, 253–254

pumped storage energy, 233

Q

Qaanaaq, Greenland, 152

quotas, for energy consumption, 180

R

radiation, 28–29

radiative forcing, 340

radicalism, as a role of NGOs, 274–275

rainfall, 111

ratification formula, 191

ratifying climate pacts, 197–198

Real Climate (blog), 294

recreation opportunities, 169

recycling, 252

Redford, Robert (celebrity), 293

reducing
  cooling in buildings, 256
  energy demand, 231–232
  heating in buildings, 256
regenerative agriculture, 25,
  267, 323
regional governments
  about, 22
  adaption and, 168–169
regulations
  about, 178
  building, 178–179
  energy use, 179–180
  self-, 179
  taxes, 180–182
relative humidity, 44
Remember icon, 3
renewable energy
  about, 223–224
  capturing carbon dioxide,
    228–231
  combining heat and power,
    226–227
  efficiency and, 224–225
  energy demand, 224–227,
    231–232
  fossil fuel emissions, 227–231
  geothermal energy, 238–239
  hydropower, 239–240
  in India, 212
  ocean power, 240–241
  options for, 232–245
  plants, 241–242
  reducing energy demand,
    231–232
  solar energy, 235–237
  sources of, 232–233
  storing carbon dioxide,
    228–229, 230–231
  supporting, 323–324
  waste, 242–245
  wind energy, 233–324

Renewable Energy Law and
  Energy Conservation Plan
  (China), 206
Report of the World Commission
  on Environment and
  Development, Our
  Common Future
  (Brundtland Report),
  213–214
reptiles, 134
research
  programs for, 164–165
  as sources for news
    stories, 287
Resilient Cities Network,
  170–171
The Right to be Cold (Watt-
  Cloutier), 332
rivers, ecosystems of, 129–130
Roberts, Julia (celebrity), 292
Rocky Mountain Institute, 256
Rotary International, 282
rules, for emissions trading, 164
ruminants, 322
run-away greenhouse effect,
  117–118
run-of-river hydropower
  plants, 239
Russia, temperate forests
  in, 266

S

Saami people, 151
Sano, Yeb (negotiator), 219
*Saving Us, A Climate Scientist's
  Case for Hope and
  Healing in a Divided World*
  (Hayhoe), 334
Scandinavia, temperate forests
  in, 266
Scheer, Hermann, 224
Scholz, Olaf (Chancellor),
  175–176
Schwarzenegger, Arnold
  (celebrity), 293, 330
A Scientific Romance (Wright), 296

scientists, as leaders in climate
  change, 330
sea levels, rising
  about, 104–106
  global warming and, 15, 17, 19
selective harvesting, 266
self-regulation, 179
sewage systems, climate change
  and, 147
Sierra Club, 273, 283
Sirota, David (screenwriter),
  289–290
skin cancer, worsened by global
  warming, 142
small modular reactors
  (SMR), 246
soil, 36, 268
solar, wind, and battery
  technologies (SWB), 178
solar cycles, as a cause of global
  warming, 9, 49
solar energy, 24, 177, 235–237
solar flaring, 340
solar photovoltaic energy, 233
solar thermal energy, 236–237
solid biofuels energy, 233
solid waste, 242–243
sources, for news stories,
  286–288
South Africa, climate initiatives
  in, 183
Spain
  nuclear power in, 247
  wind energy in, 233–234
species adaptation, 123–124
Spielberg, Steven (director), 289
Standing Rock, North
  Dakota, 277
State of Fear (Crichton), 296
states, climate initiatives in,
  183–184
steel, manufacturing, 82
Stern, Nicholas (economist), 145
Stiglitz, Joseph (economist),
  145–146

storing carbon dioxide (CO$_2$), 228–229

storms
about, 113–114
global warming and, 15

storm-warning systems, 168

subnational governments, 183

sugarcane, 242

sulfur hexafluoride (SF$_6$), 42

Sundance Film Festival, 293

sunspots, 340

surveys, as sources for news stories, 288

Sustainable Communities and Cities campaign, 167

sustainable development
about, 213–214
sustainable development goals (SDGs), 213

sustainable development goals (SDGs), 213

sustainable energy. *See* renewable energy

Sweden
carbon taxes in, 182
governments in, 176
nuclear power in, 247

Swiss RE, 263

SwissEnergy, 186

Switzerland, climate initiatives in, 185–186

T

taiga, 131

tailing ponds, 255

Taiwan, solar energy in, 237

targets, for energy consumption, 180

tax regulations, 180–182

tax shifting, 181

Technical Stuff icon, 3

Tembec, 253–254

temperate forests, 266

temperature
degrees of, 55
heat, 116–117
increased, 57–58
of oceans, 125

territories, climate initiatives in, 183–184

Texas, climate initiatives in, 183

thermal energy, 236–237

Third Pole, 107

13th Five Year Plan for Electricity (China), 206

This Changes Everything (documentary), 290

350.org, 282

Thunberg, Greta (activist), 279, 291, 333

ticks, 140

tidal energy, 177, 240

The Time Machine (Wells), 296

Tip icon, 3

tipping point, 56–57

tourism opportunities, 169

towns, climate initiatives in, 182–183

Toyota, 259

transmission lines, climate change and, 147

transportation
in China, 207
climate change and, 147
daily living and energy-efficient, 319–322
driving, 93–94
efficiency in, 225
emissions from, 92–95
flying, 94–95
governments and, 171–174
greenhouse gases (GHGs) and, 84–87

tree cover, expansion of, 168

trees, as carbon sinks, 87–88

tropical forests, 131

tropical species, endangered, 134–135

truck emissions, 85–87

Trump, Donald (US president), 170, 198

U

UN Climate Change Conference, 269

U.N. Food and Agriculture Organization (FAO), 322

U.N. Framework Convention on Climate Change (UNFCCC)
about, 23, 189–190, 193, 198
countries in, 192–193
duties of, 190
establishing game plans, 191–192

Union of Concerned Scientists, 283

United Cities Global Governance, 170

United Kingdom
as an Annex 1 country in U.N. Framework Convention on Climate Change, 192
Building Research Establishment and Environmental Assessment Method (BREEAM), 258
carbon taxes in, 182
Climate Challenge, 348
climate initiatives in, 183
government education from, 165
governments in, 177
NGOs in, 282–283
renewable energy sources in, 232

United Nations Declaration on the Rights of Indigenous People (UNDRIP), 277

About the Authors

Elizabeth E. May has been recognized internationally for her work in the environmental movement. She assisted in organizing the first international, comprehensive scientific conference into the climate change threat, in June 1988, hosted by Canada. May was Executive Director of Sierra Club of Canada for 17 years, before leaving that position in 2006 to enter politics. She is past Leader of the Green Party of Canada, having been the longest serving woman leader of any party in Canada (2006–2019). She was the first Green elected to Parliament in Canada and continues to serve as MP for Saanich-Gulf Islands. May is a lawyer, an author of nine published books and, most importantly, a mother and grandmother. Among many prestigious Canadian awards and honors, she has received the highest citizen honor in Canada, the Order of Canada, at the Officer level.

John Kidder has been a cowboy, a miner, a fisher, a range management specialist, an environmental economist, a technology entrepreneur, a small farmer, and a governance practitioner. He was a founder of the Green Party of British Columbia in 1982 and has been active in electoral politics since then. He is a past Vice President of the Green Party of Canada. Kidder holds six patents in fiber optic technology. He has a special interest in the economics and finance of transformative disruption. He married Elizabeth May on Earth Day 2019.

Dedication

Elizabeth dedicates this book to John and our children and grandchildren, in hopes that by the time the youngest of you is old enough to read this book, the prognosis will be very different and far more hopeful.

John dedicates this book to the young activists around the world who refuse to accept that his generation keeps letting them down and who continue their fierce engagement on the front lines of resistance.

And it goes without saying that they both dedicate this book to you, the reader, for making the choice to read about climate change.

Authors' Acknowledgments

John and **Elizabeth** both want to express deep appreciation to many friends and colleagues who assisted in the research and writing of this book. A special thank you to Dr. Ian Burton, Dr. Jim Bruce, and Dr. Gordon McBean, leading scientists of the Intergovernmental Panel on Climate Change, who helped ensure the accuracy of the 2009 edition. We're grateful to those that have helped review chapter content: Dr. O.W. Archibold of the University of Saskatchewan; Dr. Jonathan Newman of the University of Guelph; Peter Howard of Zerofootprint; Ruth Edwards of the Canada Climate Action Network; Kristopher Stevens of the Ontario Sustainable Energy Association; and to the David Suzuki Foundation team of Nick Heap, Paul Lingl, and Dale Marshall. (As always, any errors and omissions are the authors' alone.) Thanks also to key image providers, Dr. Max Boykoff of the Oxford University Centre for the Environment, and John Streicker of the Northern Climate Exchange.

To Debra Eindigeur, Elizabeth's Chief of Staff for assistance in endless way. To Cendrine Huemer and Jaymini Bihka for their research work. Ongoing gratitude to the countless colleagues called on for advice, feedback, or data. To Celine Bak for her dedicated work on climate disclosure in financial matters.

John and **Elizabeth** also want to express their deepest gratitude to the seemingly endless patience of our editors Robert Hickey and Chad Sievers, for always-excellent advice, text maneuvering, and overall guidance. A big thanks to those who worked behind the scenes.

Elizabeth wants to say that (once again, as in previous books) nothing would be possible without the extraordinary grace, patience, and support of her daughter, Cate May Burton. No one has ever had a better daughter, and few have known a better person.

John wants to thank his late wife Siri Heiberg, his children Janet, Kendall, and Julia and their many cousins, for their patience with his ranting over all these years and for their many insights that have helped him when he gets too academic. He thanks his father Kendall for instructing him about climate change in the 1970s, and his mother Jill for always insisting on research before writing and ensuring that the mind is engaged before putting the mouth in gear. And he thanks his dear wife Elizabeth for her tireless work to make the world a better place, her inspirational leadership, and for never ever even contemplating giving up.

This book would never have happened if not for the inspiration of Zoë Caron who initially recruited Elizabeth to be her co-author. She has gone on to incredible work for NGPs and in government.

This book was made possible by people who — intentionally or not — provided the most timely, impromptu, and gracious writing locales: the owners of Coburg Coffee in Halifax and of Planet Coffee and Bridgeheads in Ottawa, Liz McDowell, Louise Comeau, Panny Taylor, Candace Batycki, Adriane Carr and Paul George, Anjali Helferty and Roxanne Charlebois, Kathryn Kinley, and Reina Lahtinen.

Publisher's Acknowledgments

Executive Editor: Lindsay Sandman Lefevere

Project Editor: Chad R. Sievers

Technical Editor: DeWayne Cecil, PhD

Production Editor: SaiKarthick Kumarasamy

Cover Image: © Joanna McCarthy/Getty Images